INVESTMENT CASTING
HANDBOOK

Edited by

H. T. Bidwell

PRINTED IN THE UNITED STATES OF AMERICA

Preface

The prime purpose of this publication is to provide a general introduction to the investment casting process and its capabilities. It is hoped that this information will help generate awareness of the potential that the process provides in respect to freedom of design, improved design and lower component costs which are associated with this near net shape, often net shape, process.

The need to consult and collaborate with the investment casting suppliers is stressed many times. Close consultation and collaboration lead to the optimization of the benefits inherent in the process.

It is sincerely hoped that this publication will prompt designers and engineers and purchasing officers to ask themselves not whether they should use investment castings, but how best to use the process to design new parts and how to convert existing components to investment castings.

Acknowledgments

The Investment Casting Institute acknowledges the cooperation of member companies for supplying new information and examples of investment casting applications. The Institute also gratefully acknowledges those past and present Directors of the Institute who have devoted so much time and effort to the production of this publication.

CONTENTS

Chapter 1

The Investment Casting Process

Historical

The investment casting process can be authoritatively traced back to 4000 years B.C. and has been applied with particular success to the production of art castings and jewelry over the centuries. The progress of the process has been traced from China to India to Egypt to Africa and Europe. In West Africa, a large number of castings were produced from about the eleventh century onwards. During the sixteenth century, the process was widely applied by artists and sculptors. Benvenuto Cellini produced many works of art by the process; one of the most outstanding being a bronze statue of Perseus and the head of Medusa. Cellini left a graphic description of the process in a treatise which he published in 1568. Vavrinec Krickes, who lived in Prague during the mid sixteenth century, wrote "A Guide to the Casting and Preparation of Cannons, Balls, Mortars, Bells", in which he describes the use of the lost-wax process to produce the art work on bronze gun barrels. The process was also used extensively by early Colombians and by the Aztecs.

The lost-wax process is still used in statuary casting today and employs the same basic methods, although advantage is taken of modern aids. The lost-wax process was adapted by dentists during the late nineteenth century to produce accurate castings, often in gold, for fillings, crowns, and bridges to individual requirements. The metal was cast by centrifugal techniques, which assisted good reproduction of fine detail.

The continued development of the process during the early years of the twentieth century laid the foundations of the engineering process as it is known today. The need to produce accurate dental castings led to the study of factors affecting pattern and mold stability and to the solidification and contraction characteristics of a number of metals and alloys. Many techniques were introduced to obtain dimensional accuracy of the casting; particular attention was given to the mold and mold materials and a number of ingenious casting devices were developed to try to counteract expansion and contraction effects. One of the more notable developments in this field was the introduction during the early 1930's of a compensating investment material. The fact that at least four hundred patents covering the process were granted during the period 1900 to 1940 gives an indication of the interest in investment casting.

The introduction of the investment casting process to the jewelry trade occurred almost by accident, but once introduced the technique was widely applied. The process had been used on a 'one-off' basis but it was soon appreciated that castings could be mass produced if dies were used to produce the patterns in quantity and then assembled onto suitable running systems.

No serious effort was made to produce industrial castings by the investment casting process until the late 1930s, when it was realized that a cobalt-base alloy, which had been developed by the Austenal Laboratories for surgical implant work, possessed highly desirable properties at elevated temperatures. Certain components for aero-engine turbo-chargers operated under rigorous conditions and conventional alloys were found to be less than satisfactory for these applications. The cobalt-base alloy was suitable for the operating conditions but was very difficult to machine or work. There was an obvious need to develop a "cast-to-size" process to produce components in this alloy and investment casting was the inevitable choice. The potential of the process was soon appreciated and industrial techniques were rapidly developed to serve the specialized requirements of the aero-engine industry.

Most early investment foundries were tied to aircraft companies or in government establishments, but a number of commercial foundries were set up to produce aircraft quality castings. Investment casting markets gradually expanded into the field of commercial castings, and soon the engineering industry was utilizing castings produced from a wide range of ferrous, nonferrous and light alloys. The industry has developed over the years and serves a very varied market from golf club heads to turbine blades. Castings are produced in a variety of alloys in weights ranging from a few grams to many hundreds of pounds. The process is versatile and competes very effectively with other manufacturing processes, often producing a better component at less cost.

Process Description

Investment casting is an industrial process which is closely controlled at every point of production. Every casting shipped —and this applies equally to orders for a dozen castings or half a million—can be relied upon to meet the designer's performance specification.

Pattern Production

The process begins with production of a one-piece heat-disposable pattern. This pattern is usually made by injecting wax or plastic into a metal die. These dies may range from a simple, hand-operated single cavity tool to a fully automated multi-cavity tool, depending on production quantities and complexity of the part.

A heat-disposable pattern is required for each casting. These disposable patterns have the exact geometry of the required finished part with allowances made to compensate for volumetric shrinkage (a) in the pattern production stage and (b) during solidification of metal in the ceramic mold.

The pattern carries one or more gates which are usually located at the heaviest casting section. The gate has three functions:

- to attach patterns to the spruce or runner, forming a cluster;
- to provide a passage for draining out pattern material as it melts upon heating;
- to guide molten metal entering the mold cavity in the casting operation; and to ensure a sound part by feeding the casting during solidification.

Pattern Assembly

Patterns are fastened by the gate to one or more runners and the runners are attached to a pouring cup. Patterns, runners and pouring cup comprise the cluster or tree, which is needed to produce the ceramic mold. The number of runners per section and their arrangement on the pouring cup can vary considerably, depending on alloy type, size and configuration of the casting.

Figures 1, 2 and 3 show the sequence of pattern production and assembly. Figure 1 represents the injection of wax or plastic into the pattern die. Figure 2 represents the removal of the solidified pattern from the die; it is now ready to be assembled onto a cluster or tree Figure 3.

The Ceramic Shell Mold Process

This process has largely replaced the original block or solid mold process except for some smaller high volume parts and for some aluminum casting where the mold is plaster bonded.

The ceramic shell mold technique involves dipping the entire cluster into a ceramic slurry, draining it, then coating it with fine ceramic sand. After drying, this process is repeated again and again, using progressively coarser grades of ceramic material, until a self-supporting shell has been formed. The shell may be from 3/16 to 5/8 in. thick. - Figures 4, 5 and 6.

The coated cluster is then placed in a high temperature furnace or steam autoclave where the pattern melts and runs out through the gates, runners and pouring cup. This leaves a ceramic shell containing cavities of the casting shape desired together with a suitable running and feeding system, Figure 7.

Casting

The ceramic shell molds are fired to burn out the last traces of pattern material, to develop the high temperature bond of the ceramic system and to preheat the mold in preparation for casting. Because shell molds have relatively thin walls,

How It Works

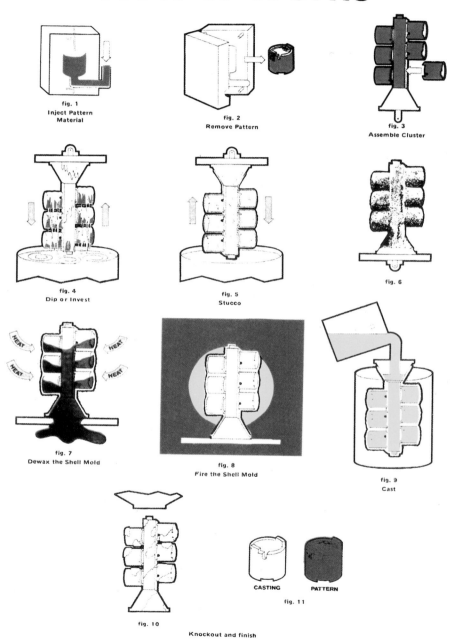

fig. 1
Inject Pattern
Material

fig. 2
Remove Pattern

fig. 3
Assemble Cluster

fig. 4
Dip or Invest

fig. 5
Stucco

fig. 6

fig. 7
Dewax the Shell Mold

fig. 8
Fire the Shell Mold

fig. 9
Cast

fig. 10

CASTING   PATTERN
fig. 11

Knockout and finish

they can be fired and be ready to pour after only a few hours in the furnace, Figure 8.

The hot molds may be poured utilizing static pressure of the molten metal heat, as is common in sand casting, or with assistance of vacuum, pressure and/or centrifugal force. This enables the investment casting foundry to reproduce the most intricate details and extremely thin walls of an original wax or plastic pattern, Figure 9.

Melting equipment employed depends on the alloy. For nonferrous alloys, gas fired or electric crucible furnaces are usually used. High-frequency induction furnaces are most commonly used for melting ferrous alloys.

Cleaning

After the poured molds have cooled, the mold material is removed from the casting cluster. This is done by mechanical vibration and chemical cleaning. Individual castings are then removed from the cluster by means of cut-off wheels and any remaining protrusions left by gates or runners are removed by belt grinding. The casting is then ready for secondary operations such as heat treating, straightening, machining and for whatever inspection is specified.

Chapter 2

How to Buy Investment Castings

Why Choose an Investment Casting?

The investment casting process competes with the majority of metal forming processes and offers some unique advantages. The process offers a multitude of benefits including cost savings, design freedom, close tolerances, better finishes, savings in machining time, reproducibility, and assembly savings.

Low initial tooling costs: Initial tooling costs are lower than for most other metal forming techniques. Although most tooling is of aluminum alloy, prototypes can be made by using epoxy or aluminum filled epoxy materials. New techniques in rapid prototyping can provide prototype patterns where needed for evaluation purposes. Tooling costs, averaged over the parts produced, are often lower than other manufacturing tooling costs.

Elimination of material waste: Investment castings are essentially cast to size. As a result there is little machining necessary with consequent saving in time and material costs.

Design flexibility and capability: The investment casting process offers infinite alloy choices, and unlimited design flexibility for external and internal configurations.

Design Enhancements: Unlike other casting methods, there is no draft requirement in the investment casting process .

Consistency: The investment casting process gives a reliable and consistent product. Due to positive process control programs, remarkable consistency is maintained within batches and from batch to batch.

Close Tolerances: Investment casting produces the closest tolerances of any casting process over a wide range of alloys.

Surface Finish Improvement: A surface finish of about 125 RMS (root means square) is typical for steel castings. No casting process produces a finer surface finish than the investment casting process.

Where to Find an Investment Caster

Currently there are approximately 350 domestic sources for investment castings. Typically, names and addresses can be found in industrial journals and magazines or by attending local trade shows. The simplest way to locate an investment caster is to contact the Investment Casting Institute. A listing of all member foundries will be provided which will include names, addresses, telephone numbers, alloys poured, and size ranges of castings. This listing will allow the casting buyer to contact those investment foundries that may realistically meet the buyer's needs.

Many Investment Casters have areas of specialization. Some are predominantly aerospace, while others lean more towards commercial grade castings. It is best to ask the foundry just what type of foundry they are, whether or not they can pour the alloys that you require, and whether or not your quality requirements can be achieved.

It is strongly recommended that you take the time to visit the investment foundry prior to placing a purchase order. Get to know the foundry's capabilities and capacity, meet the people that you will be dealing with, the engineers, quality control, and production people who actually get the job done.

When to Contact an Investment Caster

Make the Investment Caster part of your design team. His knowledge of the metal processing industry will enable you to choose the right process to produce your parts in the fastest, most cost effective manner.

It is best to contact the Investment Caster as early as possible in the design phase of a new project. The foundry engineers may be very helpful in reducing cast weight (and cost), improving strength, selecting alloys, and suggesting design alternatives.

Often, simple, non-functional design changes can be incorporated which will drastically reduce tooling costs.

One of the most attractive aspects of the investment casting process is the versatility of design. The casting buyer should take full advantage of the expertise offered by the foundry's engineers. This will not only ensure the lowest possible product cost, but will help to prevent costly delays due to re-engineering.

Quotations

When issuing a Request for Quotation, whether it be to one or several foundries, be as specific as possible. Clearly state the alloy or alloys to be cast, any machining or secondary operations needed, and any NDT (non-destructive testing) or material certification requirements. If volume or cast weight is available, please include this information.

When requesting various quantity breaks, please be realistic. State the number of castings normally required per order and the annual requirements. The type of tooling quoted is very dependent on the quantities requested. Automatic or multi-cavity tooling may reduce individual piece price, but initial tooling costs will be much higher than for a single cavity, manual tool. Request that the investment caster describe the type of tooling quoted.

Ask the casting engineers to review all drawings and clearly state any exceptions or suggestions that they might have.

Be sure the foundry knows your lead time requirements as these can have an important effect on the foundry's quote. Lead times stated on the quotation are usually based on a "best case" scenario; assuming that the samples are inspected and approved promptly.

The investment caster should clearly state on the quotation all payments terms and conditions. Any special arrangements that you might require should be discussed at this time.

The investment caster's sales department may be responsible for developing the quotation, but very often they will seek input from quality control, production, and engineering departments. Their goal is to provide you with the most accurate quotation possible, and in a timely manner.

Blueprints and Drawings

Probably the most important document in any investment foundry quoting effort is the part drawing The more information that is on the drawing, the less chance that there will be for a misunderstanding. Although facsimile copies may be suitable for quotation purposes, it is not recommended that they be used to build tooling and produce castings. Very often some portions of these copies will be illegible. A hard copy of drawings should be provided whenever possible.

All drawings should clearly state the following information:

PART NUMBER
PART NAME
REVISION LEVEL
ALLOY(S)
TOLERANCES
FINISH REQUIREMENTS
CRITICAL DIMENSIONS
MACHINING REFERENCE POINTS (if applicable)
INSPECTION REQUIREMENTS
HEAT TREAT CONDITION

It is strongly recommended that separate drawings be developed for castings and machining. The investment caster's quality control department and your incoming inspection department should be provided with drawings showing exactly how the castings are to be received at your dock. This will eliminate confusion and possibly costly delays.

The Purchase Order

The purchase order is a legally binding contract between the customer and the investment caster. As with any legal document, the more specific it is, the better. The purchase order should include all of the following information:

PART NUMBER
DRAWING NUMBER
REVISION LEVEL
QUANTITY ORDERED
PURCHASE ORDER NUMBER
PIECE PRICE
DATE REQUIRED
BUYER'S NAME
SHIP TO ADDRESS
BILL TO ADDRESS
MATERIAL CERTIFICATION REQUIRED
HEAT TREAT CONDITION REQUIRED
ANY OTHER CERTIFICATION OR TESTING REQUIRED

All purchase orders should require that the order be acknowledged in writing by the investment foundry. It is the responsibility of the investment caster to inform you if any conditions of the purchase order are unacceptable. The customer should review the acknowledgment and clarify any areas of doubt before production is approved.

Tooling

At the time of the initial quotation, you will be quoted a piece price cost and a tooling cost. Most foundries will quote tooling as a separate item. This is usually a one time cost. It is wise to request that the investment caster describe the type of tooling they are quoting, whether manual, automatic, single or multi-cavity. Tooling is usually considered to be the property of the customer after all outstanding invoices are paid. Tooling is normally stored and maintained in an operational condition by the investment caster. It is critical that the issues of ownership and maintenance be clearly established at the beginning.

Investment casting tools are readily transferable from one foundry to another, often with little or no re-rigging costs. Some foundries however, have specialized wax injection equipment which may or may not be compatible with all types of tooling. For instance, a foundry which operates only manual wax injection machines may require some tool modifications in order to operate automatic tools. Again, it is advisable to discuss this with the investment foundry in advance of relocating tooling and determine whose responsibility it would be to pay for any modifications. Please remember that it will be very difficult for the investment foundry to determine exact modification costs until they have physically examined the tooling in question.

Samples

Producing several sample castings prior to full scale production serves a dual purpose. First, the parts must be checked to the drawing to assure dimensional accuracy. Often the caster's Quality Control department will inspect the tooling and several wax samples and calculate appropriate shrink factors to determine dimensions. In the case of relocating tooling from one foundry to another, this inspection procedure must still occur. Differences from one company to another in the type of wax used, shell building methods, and pouring practices may cause variations in shrink factors. Usually these differences will be minimal, but on critical dimensions they may be great enough to warrant a tool modification. Most investment casters will gladly tell you to what shrink factor your tooling was built. It is advisable to keep this information with your pattern records for future reference. In the event that tooling must be relocated at a later date, you can inform the new foundry exactly what type of tooling they will be receiving.

The second reason for running sample parts is to prove out the process engineering. Through continual process improvement, investment casters are constantly searching for ways to reduce their manufacturing costs. In the initial sampling process, the foundry may attempt several engineering methods to achieve the highest quality at the lowest cost. Even if tooling has been relocated and shrink factors determined, the investment caster will require that samples be produced and approved prior to releasing production orders. Wax injection methods, ceramic shell procedures, and metal temperature, will all need to be determined and monitored by foundry engineers.

The sampling stage is a critical period in all new product development, and proper communication and cooperation between the buyer and the investment caster is critical to ensure that proper quality and delivery requirements are met.

Chapter 3

Dimensions, Tolerances, Surface Texture

The cost of any casting increases in proportion to the restrictiveness of the specifications for dimensional tolerances, chemistry or nondestructive testing.

The customer should be very aware that close consultation with the engineering staff of the investment caster can often lead to better design and to a lower cost product.

Tolerances

Tolerances may be affected by a number of variables. Wax or plastic temperature, injection pressure, die temperature, mold or shell composition, back up sand, firing temperature, rate of cooling, position of the part on the "tree", and heat treat temperature -- all bear directly on tolerances required in the investment casting industry. The amount of tolerance required to cover each process step is dependent, basically, on the size and shape of the casting and will vary from foundry to foundry. This is because one foundry may specialize in thin walled, highly sophisticated castings, another in mass production requirements, and yet another in high integrity aerospace or aircraft applications.

(a)   **Linear tolerancing** is normally applied to the following features of investment castings:

Length	Flatness
Concentricity	Straightness
Fillet radii	Corner radii
Holes	Curved holes

(b)   **Geometric Tolerancing** is normally applied to the following features of investment casting:

Profiles & true positioning	Roundness
Parallelism	Perpendicularity
Contours	Tapered holes
Radii	Cam profiles

Linear Tolerancing

As a general rule normal linear tolerances on investment castings are as follows: Up to 1" ± .010", for each additional inch up to ten inches ± .003" per inch. For dimensions greater than ten inches allow ± .005" per inch. Secondary operations such as straightening and sizing will produce closer dimensional tolerancing.

LINEAR TOLERANCE	
DIMENSIONS	*NORMAL*
up to 1"	± .010"
up to 2"	± .013"
up to 3"	± .016"
up to 4"	± .019"
up to 5"	± .022"
up to 6"	± .025"
up to 7"	± .028"
up to 8"	± .031"
up to 9"	± .034"
up to 10"	± .037"
> 10" allow ± .005" per inch	

An exception to the Linear Tolerance exists on wall thickness where the tolerance must be a minimum of ± .020"

It should also be understood that in applications such as gas turbines and defense oriented products, very much closer tolerances are routinely achieved in the as-cast state.

Flatness

Flatness and straightness are so closely related that confusion may arise unless the foundry and the purchaser reach definite agreement prior to production. Flatness tolerance is the total deviation permitted from a plane and consists of the distance between two parallel planes between which the entire surface so toleranced must lie. In measuring, the parallel planes must be the minimum distance apart.

The degree of flatness exhibited in an investment casting is almost always determined by the amount of volumetric shrinkage that the wax and metal undergo during cooling. This shrinkage is usually in the center of the mass and is referred to as "dish" (shrinkage, dip, or "out of flat"). This dish can be controlled (premium) by specialized techniques, but will always be present to some extent. General flatness tolerances cannot be quoted as they vary with configuration and alloy used.

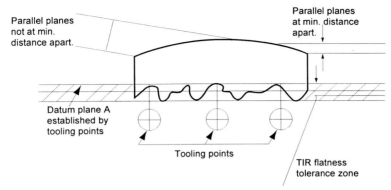

Straightness

The tolerance covering the straightness of an axis is the diameter or width within which the axis must lie.

It is obvious from this that to correctly measure axial straightness of either a shaft, bar or plate, the tolerance zone (within which the axis or axial plane lies) must be measured.

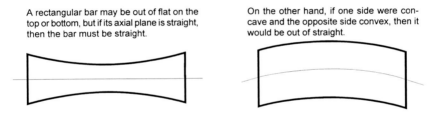

A rectangular bar may be out of flat on the top or bottom, but if its axial plane is straight, then the bar must be straight.

On the other hand, if one side were concave and the opposite side convex, then it would be out of straight.

Straightness may be a real problem with certain types of castings. A relatively thin, short part may bend while a long heavy part may not. Experience tells the foundry that a given design may bend, but experience cannot say to what extent. As a rough guide, it may be said that a constant section will have an axial bow of 0.005" per inch. Ribs and gussets will inhibit warpage and will also hinder the mechanical straightening.

Parallelism

Casting of parts, which have parallel prongs supported only at one end, present a very specialized type of problem and should be discussed fully with the foundry prior to production.

13

Yoke castings also present a very specialized type of problem and should be discussed fully with the foundry prior to production.

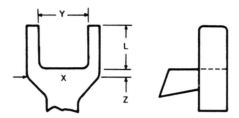

Since point X is the thickest section, it is the ideal point to gate. It is also the area where the greatest volumetric shrinkage will occur. Dimension Y, however, will be restrained by the rigid mass of refractory. The result is that parallelism is difficult to maintain and will be 0.010" per inch of L, but can be improved by control techniques and sizing. This condition will also affect any through holes usually found in yokes. When specified, such holes should carry considerable finish stock if they are to be finished truly concentric or line reamed.

Roundness or "out of round"

"Out of round" is defined as the radial difference between a true circle and a given circumference. It is the total indicator reading when the part is rotated 360° or it can be calculated by taking half the difference between the maximum and minimum condition The latter technique is usually preferred since it takes less time. The actual method of inspection to be used, however, should be specified by the purchaser.

OUT OF ROUNDNESS	
Diameter	TIR or ½ difference between diameters
½"	.010"
1"	.015"
1½"	.020"
2"	.025"
On larger diameters, linear tolerances apply	

Concentricity

Two cylindrical surfaces sharing a common point or axis as their center are *concentric*. Any dimensional difference in the location of one center with respect to the other is the extent of *eccentricity*.

The sketch shows that out of roundness in either diameter does not affect concentricity because concentricity relates the centers or axes of the diameters. Out of roundness is their variance from a true circle.

14

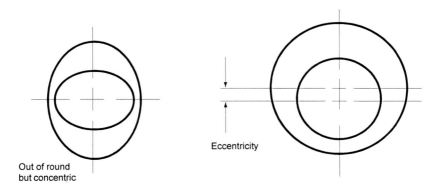

Out of round
but concentric

Eccentricity

However, in a shaft or tube, straightness has a very real influence on con-centricity.

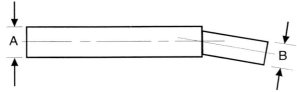

Diameters A and B may be true circles, but it is obvious that the out of straightness condition has affected concentricity.

When the length of a bar or tube does not exceed its component diameters by a factor of more than two times, the component diameters will be concentric within 0.005" per inch of separation.

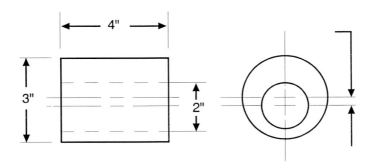

EXAMPLE A - 3" OD x 2" ID x 4" long. 3" OD and 2" ID will be concentric within 0.005" TIR (3" OD - 2" ID = 1" separation)

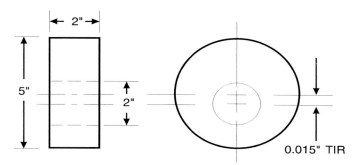

EXAMPLE B - 5" OD x 2" ID x 2" long. 5" OD and 2" ID will be concentric within 0.015" TIR (5" OD - 2" ID = 3" separation)

When the length exceeds the factor of two times, then the amount of out of straightness as described in Examples A and B should be added to the inherent eccentricity.

EXAMPLE:
2" OD x 1" ID x 4" long. Separation = 1", eccentricity .... = .005" TIR
4" X .005" per inch out of straightness .... = .020" TIR
Total of deviation .... = .025" TIR

Angularity

Angular tolerance is dependent on the configuration forming the angle.

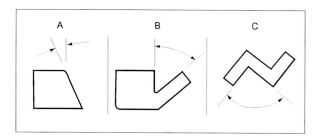

Sketch "A" cannot be sized, but in certain cases after sufficient data has been reviewed, the die can be reworked to bring the part closer to nominal dimension. Sketches "B" and "C" can be reworked to ± ½° and ± 1° respectively. Obviously, however this is dependent on the alloy and its condition.

Hole Tolerance

The roundness of a cast hole is affected by the mass of surrounding metal.If an uneven mass is adjacent, the hole will be pulled out of round.If the surrounding metal is symmetrical, holes up to ½" diameter can be held to ± 0.003" when checked with a plug gage. Larger holes may be affected by interior shrinkage or pulling, and the foundry should be consulted.

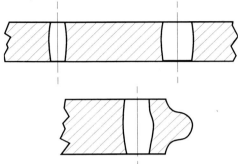

The longer the hole or the more mass of the section around it, the more pronounced the effect. Some shrinkage concavity will be present to some extent in all castings. The openings at top and bottom of the hole will be approximately the same dimension while the center will be a larger diameter. Through holes which require clearance (this can be checked using a plug-type gage) can be held to fairly close tolerances if the larger diameter in the center is ignored. If, however, the sidewalls of the hole are used as bearing surfaces, a simple reaming operation will size the cast opening.

The lower figure shows the effect of shrinkage on a hole diameter when a heavier section is in the proximity of the hole itself. Note that the diameter is distorted due to additional mass shrinkage of the heavier section. The figure is a graphic illustration of the distortion which will be present to a greater or lesser degree in every casting when a heavier mass affects shrinkage.

Curved Holes

Since curved holes are formed by either soluble wax or preformed ceramic cores, the normal tolerance tends to be doubled. A factor of two times must be applied to the tolerance on all dimensions controlling such a feature. Since such holes cannot be sized, a diameter tolerance of ± 0.005" per inch also applies.

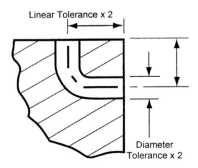

Linear Tolerance x 2

Diameter
Tolerance x 2

Angular Holes

Since these holes are usually formed by metal cores within the die, the tolerance restrictions for curved holes do not apply and normal tolerance bands are usually acceptable. If the angle formed by the two centerlines is greater than 120°, the hole can be sized, but if it is less, a diameter tolerance of ± .005" per inch must be used.

Internal Radii, Fillets

These should always be given as wide a tolerance as possible. They are difficult to control and can only be checked approximately by radius gages, or at a premium by an optical comparator.

Surface Texture

In general, industry uses the term RMS (Root Mean Square) and AA (Arithmetic Average) interchangeably. RMS gives about 11% higher values than AA but the difference is not significant; the investment casting industry will measure to either standard. In order to correspond with the published literature of Investment Casting Institute member companies, the values expressed in this section are shown in terms of both RMS and AA. Both are measured by using a profilometer.

The surface texture standards for investment castings may be expressed by any of the following methods: 125 RMS max. or 115 AA max., All lay is multidirectional except in ground areas. Secondary operations will improve the surface texture to give results corresponding to the equivalent wrought alloy.

Drawing Notes

To assist the foundry in making an accurate cost estimate of a casting, certain information should be on the drawing in the form of notes. This is also necessary during production, while dies and tooling are being made, during casting and afterwards, for casting inspection. The following notes should be on every casting drawing:

Notes concerning geometry of the casting

Note 1: Tolerances, flatness, perpendicularity, parallelism and concentricity.
Tolerances are sometimes placed in the drawing box, sometimes in the form of a drawing note. Larger companies often spell them out in a general engineering specification. The linear tolerances, however, should also be marked on the drawing.
Example: "Linear tolerances, unless otherwise specified, ± .010" for first inch plus ± 0.003" per inch for every inch thereafter." If you do not spell out your requirements for linear tolerances, flatness, perpendicularity, parallelism and concentricity, the investment casting foundry will produce the castings to standards set forth in this chapter.

Note 2: Fillets and corner radii
Incorporating corner radii in investment casting boosts tooling costs, since multiple cavity work becomes necessary. Specify these radii only if they fulfill a technical or aesthetic function that is worth the extra cost. Cast corners of investment castings always have a natural radius of about 9.003" up to 0.010". Corners generated by machining tend to be sharp and a separate operation may be needed to break them.
Specify, as "max," all corner radii when possible. Fillet radii, on the other hand, do not increase tooling costs greatly. They improve quality and mechanical strength of the casting appreciably by eliminating shrinkage (which tends to develop along sharp inside corners during solidification). Always specify fillet radii, whenever design permits this.
Example: "Corner radii 0.030" max.; fillet radii 0.060" max. unless otherwise specified."

Note 3: Thickness of walls and ribs
To save time in drafting, typical wall thickness and rib width are sometimes taken into drawing notes. Specify the broadest possible tolerances.
Example: "Wall thickness
$0.080" \; {}^{+\,0.015"}_{-\,0.000"} \;$ typical."

Note 4: Type of casting identification
You must specify the type of identification to appear on the casting. This point is often overlooked or not clearly defined, resulting in confusion and lost time. Keep in mind that a wide variety of cast-in identification can be carried on an investment casting (See Chapter 3, Section 2, "Lettering"). Make sure to specify the type of characters to be used and their actual height and depth.
Example: "Cast part number 1/8" size characters x 0.020" deep, raised on depressed pad at location shown."

Note 5: Surface Finish
Spell this out using RMS or AA numbers, generally not lower than 125 RMS max. If figures lower than 125 RMS are shown on the drawing, the foundry automatically assumes that machining stock is to be added to the marked surface.
Example: "Surface finish 125 RMS"

Note 6: Identification of machining stock
Sometimes it is not clear from the drawing where the customer expects machining stock on the casting. Spell this out clearly to avoid delay in delivery of samples and production parts.
Example: "Add 0.040 to 0.060" stock to surfaces identified by machining mark.

Notes concerning material specifications, including heat treatment

Note 7: Material specification

It is extremely important to specify the material clearly on the drawing. Definition of the metal sometimes has a considerable influence on price of the finished casting. **Example:** "Material - aluminum alloy, MIL-C-11866-T6 Comp. 2 (356-T6)."

Note 8: Heat treating specification

Usually, the material specification also fixes the heat treating conditions for the material. But sometimes you will want a different heat treatment for further processing and this should be clearly spelled out on the drawing.

Example: "In case of weld repair, anneal casting at 1525° F for two hours and cool at 50° F per hour max to 1200° F before shipment."

Notes concerning inspection

Note 9: Inspection specification

The note specifying inspection procedure - together with material specification - is absolutely necessary for price calculation and proper processing of the casting. This fact is often overlooked. When quoting, the foundry has to assume that some inspection requirement applies, but this may turn out to be very different from the customer's requirements. Misunderstanding and delays are the result.

Notes concerning final finish of part

Note 10: Final finish of part.

Many times it helps the investment caster to know how the final part will be finished. He can then alert his inspectors to bear this in mind as they comply with the specifications for finish

Example: "Anodize per instruction attached."

Chapter 4

Design and Application of Investment Casting

The design freedom and the scope of applications associated with the investment casting process cannot fail but impress design engineers seeking better and more cost effective manufacturing methods. This chapter on design and applications of investment castings presents some basic casting design concepts, a review of cast-in features offered by the process and a series of case histories and applications which should clearly illustrate the attractiveness of this process.

To obtain maximum benefit from investment castings, close cooperation between the designer and Investment Caster is essential. It is a question not merely of the best way of producing a given design as an investment casting, but of collaboration in depth so that the designer may take full advantage of the investment casting process. A component is often designed in terms of the more conventional manufacturing processes and only later is thought given to producing this design as an investment casting. It is quite possible that the design may be radically improved by the design freedom offered by this accurate manufacturing process.

Basics of Casting Design

The investment casting process is above all a casting process, and the fundamentals of directional solidification are as significant as in any other casting process. In simplest terms, molten metal should solidify in a progressive manner from the extremities of the mold towards the ingate or feeder. Such a situation ensures that molten metal is always available to feed the casting as solidification proceeds, any shrinkage contraction being confined to the feeder system.

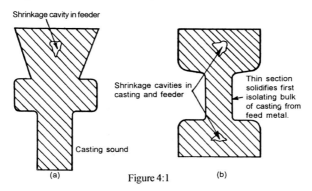

Figure 4:1

In order to maintain directional solidification and casting soundness, the running system should be designed so that thick sections feed thin sections and not the other way round. Consider the theoretical casting shown in Figure 4:1a. It is obvious that the thin section will solidify first and solidification will proceed to the boss which can be fed by a suitable feeder. If the casting were inverted (Figure 4:1b), the thin section would again solidify first, and thus isolate the boss from the feed metal.

Castings are often of complex geometry, and in order to establish satisfactory thermal gradients in the mold, the investment founder must exercise his skill by arranging the castings and feeding configuration in a satisfactory manner. Long before this stage is reached, however, attention to the design of the casting can ease the production process, minimizing development and scrap costs. The following are some of the more basic design rules:

(a) Keep section thicknesses constant.
(b) Where a change in section cannot be avoided, ensure that the change is gradual.
(c) Make use of adequate fillets so as to avoid sharp intersections, and avoid sharp exterior edges.
(d) Do not cause a number of sections to meet at the same point.

It is very seldom that a casting is designed which conforms to all these requirements, but if they are recognized a design will be produced which will offer the best possible conditions for a given component.

Dealing with these points in more detail:

(a) Uniform casting section

Consider the casting design shown in Figure 4:2a and b. The design shown in Figure 4:2a creates a number of variations in section which can lead to poor feeding conditions and the generation of hot spots. A simple change in design will produce a more satisfactory situation (Figure 4:2b).

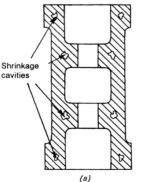
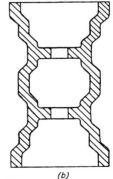

Shrinkage
cavities

(a) (b)

Figure 4:2

22

(b) Section change

The difference in thickness between adjoining sections should be kept to a minimum. Where section variations are less than 2:1, the section should be designed in the manner shown in Figure 4:3a. If the section variation is greater than 2:1, the wedge configuration shown in Figure 4:3b should be adopted, provided the section change does not exceed 4:1.

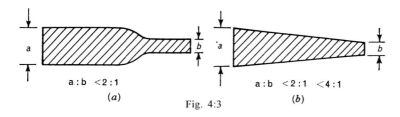

Fig. 4:3

(c) Use fillets and radii where possible

Sharp corners and intersections lead to unsound conditions and poor structural characteristics. Hot spots and severe shrinkage can be caused by badly radiused intersections (Figure 4:4a). The size of radius selected to avoid sharp corners and intersections is important. If the radius is too large, hot spots and shrinkage cavities can be produced which may be as bad as a design with no radii. As a general guide, the radius should be in the range of one-third to one-half the section thickness. Examples of good design are shown in Figure 4:4b.

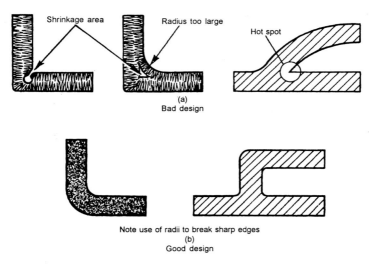

Figure 4:4

23

(d) Intersections

In view of the need whenever possible to obtain a uniform wall thickness and avoid hot spots, the problem of intersecting sections is self-evident. Consider the various sections shown in Figure 4.5a, b and c.

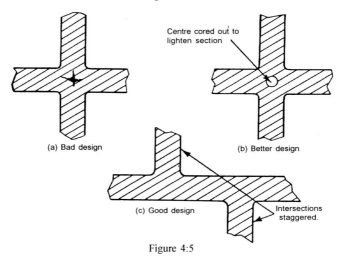

Figure 4:5

It is common practice to include rib reinforcement on castings which require extra strength or flatness. The simple cruciform intersection can lead to severe unsoundness and the ribs should be staggered to produce sections which are as uniform as possible.

The points illustrated in the general principles of design should not be underestimated since attention to such elementary considerations can ease production problems and keep the cost of the casting as low as possible.

Once the basic considerations are understood, the designer will want to know what are the limitations of investment casting design. Surface finish, limitations on the complexity of exterior and interior surfaces, what degree of straightness, flatness, concentricity can be obtained, can threads be cast, how deep a hole, how thin a section? An appreciation of these factors is important if the maximum benefit is to be gained from the investment casting process.

Cast-in Features

The uniqueness of the investment casting process allows so many features to be included in the as-cast state. Holes of all shapes and sizes; lettering both raised and depressed present no problems. Cams and slideways, simple and complex, are features commonly produced. Virtually any geometry can be readily obtained within the general tolerances described in the chapter on dimensions. A selection of some cast-in features is presented in Figures 4:6.

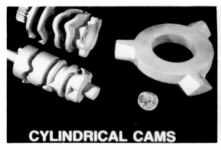

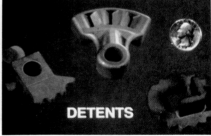

Figure 4:6

25

Directional Solidification and Single Crystal Casting

In recent years control of solidification parameters has resulted in directionally solidified castings and ultimately to directionally solidified single crystal castings.

The transitions from equiaxed grains to directionally solidified columnar grained castings to single crystals is clearly illustrated in Figure 4:7.

Figure: 4:7

Metal Matrix Composite

Yet another significant development has been the use of the investment casting process to produce castings using metal matrix composites of various types. The component in Figure 4:8 is a casting produced in an aluminum ceramic reinforced composite. Figure 4:9 shows a number of castings produced in a nickel-nickel aluminide alloy. The unique properties of these systems are invaluable for replacing more usual alloys whose chemical, physical and mechanical properties are not acceptable for advanced applications.

26

Figure: 4:8

Figure 4:9

Rapid Prototyping Options

Several rapid prototyping processes may be utilized by investment casters. Rapid prototyping systems normally generate models of components to be produced and these models allow designs to be reviewed prior to the production of the actual components.

Rapid prototyping of wax or plastic patterns for investment cast components allow the production of prototype castings without the need to produce a pattern die. Data in 'CAD' (Computer Aided Design) files is used by the rapid prototyping equipment to generate a pattern suitable for investment casting. One rapid prototyping system takes the concept one step further and produces a ceramic shell mold from design data.

The rapid prototyping concept is even being used to produce short run orders where time is critical.

The process is reliable and is used where the economics of the systems can be justified.

An example of the use of rapid prototyping is the production of tooling to be used either for the production of wax patterns or injection molded parts. Figure 4:10 shows rapid prototype patterns and the cast tool steel die halves produced from them.

Figure 4:10

Hot Isostatic Pressing

Designers should be aware of the fact that investment castings can be subjected to hot isostatic pressing (HIP) in order to achieve the ultimate in the mechanical properties. The success of the process as an adjunct to the casting process has made possible the use of large structural castings in titanium and nickel base alloys for aero engine and airframe applications.

Hot isostatic pressing is a process which exposes castings to high pressures and temperatures to minimize or eliminate internal porosity in the casting. Castings are loaded into a vessel, often on specifically designed racks so as to minimize distortion during the processing cycle. The pressure vessel is closed and purged of air with an inert gas such as argon. The castings are then heated under inert gas pressures as high as 20 t.s.i. (3000 bar) at temperatures that can be in excess of 3600°F (2000°C) depending upon the alloy. The internal porosity of the casting closes up under the effect of the high pressure and temperature and can be eliminated completely. HIP will not close up surface defects, nor will it close up internal porosity connected to the surface since the gas pressure will then be equal within the pores and on the casting surface; hence no load will be applied.

The HIP process improves the casting integrity and as a result dramatically improves mechanical properties. HIPping is already specified for a wide range of high duty castings and has been adapted to produce a lower cost but still effective secondary operation for commercial light alloy castings. Because of the 'healing' action of HIPping, it has proved possible to use less complex gating systems and hence produce less expensive castings even including the cost of HIPping. Figure 4:11 demonstrates how an aluminum alloy test casting with a high degree of porosity would be made sound by hot isostatic pressing.

Figure 4:11

Case Histories and Examples

This section presents a number of examples of castings produced by the investment casting process. The aim to illustrate the almost unlimited ability of the process to produce more detailed, more accurate, more reproducible and more cost-efficient components. It is hoped that the visual impact of these examples will encourage design engineers to consider the investment casting process as a distinct and viable manufacturing alternative to the so-called traditional methods.

The process offers true design flexibility and component reproducibility. It has already been stressed in previous chapters that one of the most critical steps engineers can take is to establish communication with investment casting companies and seriously review the advantages of the process. When comparing manufacturing methods, compare apples with apples. Investment casting should be able to produce an almost finished component and any comparison must relate to costs at the same stage of manufacture. It is often surprising how even a simple component can be produced as an investment casting less expensively, and with improved characteristics.

Welded, heavily machined component is replaced by an investment casting

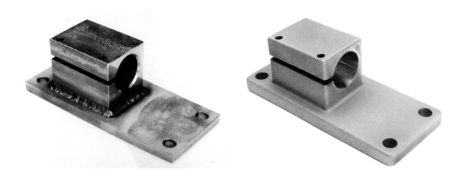

A welded, heavily machined component is replaced by an investment casting with all the features cast in to produce a less expensive, totally functional and more aesthetically pleasing end product.

Investment Cast Ejector Saves 80%; Eliminates Forming, Welding, Machining Operations

A cost savings of more than 80% resulted from investment casting this CF-3 (304 L) stainless steel net shape part, due to the elimination of expensive forming, welding and machining operations. The part represents a redesign of a five-piece welded assembly consisting of a custom-formed sheet metal box, a bulge-formed, thin-walled tube, a custom-bent, thick-wall tube, and two machined components in 304 L stainless. The center nozzle section of this casting contains two undercut features, requiring some innovative casting techniques.

Secondary operations included vacuum solution anneal heat treatment and machining of an NPT thread. As-cast tolerances of ±.005 are maintained throughout this .66-lb casting. Of course, dimensional repeatability is critical. To attain the dimensional reproducibility, the manufacturing process is monitored by statistical process control (SPC).

Five-piece
welded assembly

Investment
casting

Design & Mechanical Engineering Investment in Tooling Pays off as Investment Casting Replaces Fabrication & Weldment

The considerable design and mechanical engineering invested in a new investment cast steam turbine were well worth the effort. The new casting represents a better, stronger, lighter part, for less money.

Formerly produced as a multi-piece fabrication and weldment, the steam turbine has been produced in three sizes using the investment casting process. The required alloy had to have the necessary strength and be corrosion resistant, so the choice of 17-4 PH was not difficult. However, the tool had to be designed and manufactured so that all the critical angles and tolerances were within specifications. The physical configuration of these parts, including necessary coring, make them ideal investment castings, and the only practical method for production.

Single Investment Casting
Replaces 13 Pieces in Steering
Head of Motorcycle Chassis

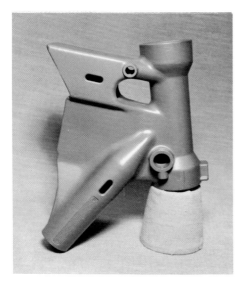 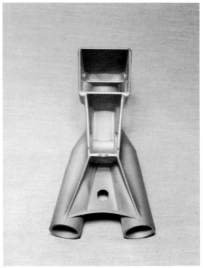

This single-piece investment cast steering head for a motorcycle chassis replaced 13 individual pieces and greatly reduced the amount of welding needed to perform the same functions in previous chassis assemblies.

The precision and accuracy of the casting eased the assembly operation and reduced the amount of rework and straightening required.

The angular and positional relationship of the chassis members are now incorporated in a single, rigid component, rather than being dependent upon complex and multiple fixturing during sequential welding operations. The steering stop-boss, fuel tank mount and a mounting hole for an optional engine guard accessory are included in the casting. Precision investment casting allowed the inclusion of a steering fork lock whose cylinder operates in a cast-to-size socket and requires only the tapping of a cast-to-size cross hole for a set screw to hold it in place.

Another feature made available by the ability of investment casting to reproduce small details and achieve fine surface finish is the tamper proof Vehicle Identification Number (VIN) pad. The $2^3/_4$"x $^{13}/_{32}$" rectangular pad has diagonal ribs raised .008" from the surrounding surface. The VIN is roll stamped into this ribbed surface as the final step of chassis fabrication. Any attempt to alter the VIN would be apparent by the inability to recreate the raised ribs.

Investment Cast Stainless Steel Bracket Reduces Price; Eliminates Assembly, Welding, Machining

By replacing a three-piece welded assembly with a single investment casting, this part consistently meets reliability requirements, at a considerably reduced price.

The investment cast bracket is used on conveyer systems made for various electronic and clean-room environments. It serves to attach various electrical, pneumatic and fluid connectors.

The new investment cast part is used as cast; without machining or other secondary operations. The part is made of 17-4PH precipitation hardening stainless steel to provide corrosion resistance and adequate strength.

Processing is strictly monitored to obtain tolerances of ±0.003" between the legs and ±0.005" on the holes that go through the legs.

Before   *After*

Conversion from Four-piece Fabrication to Single Investment Casting Saves 60% in Production Cost

When an Investment Caster quoted a part which was previously fabricated from four pieces of metal at a 60% savings, the end user quickly contracted with the foundry to produce the part as an investment casting.

The flame igniter tube is a component part of an ignition electrode used in the baking industry. Prior to switching to investment casting, a one-inch piece of stainless tubing was welded to a custom-sized screw-machined washer, and then to a $\frac{5}{8}$" piece of stainless tubing; a small piece of metal perpendicular to the tubing was then attached to the smaller tubing, which was cut away one inch from the top. The resulting part cost $11, required purchases from multiple vendors, occupied two or three departments, including purchasing. Although functional, the finished part had its sleek stainless finish marred by weld points.

When produced as a casting, the flame igniter was a smooth, solid piece of 304 stainless and cost only $6.57 when ordered in quantities of 1,000-- the company's expected annual use of the part. Annual savings for casting the flame igniter were expected to exceed $4,500.

Original part contained multiple weld points.   *Single-piece investment casting.*

Investment Casting Design Achieves Contours, Aesthetic Qualities Not Practical by Machining

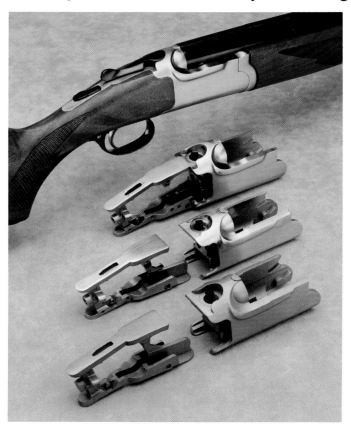

This shotgun receiver and tang is used in the Ruger Red Label Over and Under Shotgun. The assembly, made of chrome hardenable stainless steel (alloy 410), was designed as an investment casting. Machining is limited, and the pleasing contours would have been impractical if produced in any other way.

Manufacturers knew from the beginning that the investment casting process would be used on the receiver and tang. With this in mind, it was possible to incorporate superior features in the design of the firearm. The tang is cast in one piece, avoiding the more costly assemblies of less modern designs. Dimensional controls are built into the process at all stages of production. Hot straightening is used to improve final dimensional conformity. Two castings are machined prior to joining by TIG welding.

Quality, Complexity, Characterize Investment Cast Display Chassis, which Replaced Two Heavily Machined Sand Castings on M1A2 Abrams Tank

Quality and complexity characterize this investment-cast commander's integrated display chassis used the M1A2 Abrams Tank. This critical application exemplifies how investment casting technology can be used to manufacture net and near net shapes in extremely complex configurations.

Designed as a one-piece investment casting, the part replaced two sand castings and reduced the amount of machining and assembly previously required. One-piece construction also enhanced reliability and performance.

The alloy used is A356-T6, IAW MIL-A-21180, class 11. Minimum mechanical properties of 33,000 tensile, 27,000 yield and 3% elongation were determined from integrally cast specimens. Metallurgical quality requirements include a combination of grade C and D areas IAW MIL-STD-2175, of which the grade C areas are subjected to radiographic examination. Penetrant inspection is performed IAW MIL-STD-6866, Type 1, Method A on every part. There were 1191 dimensions or features requiring inspection for first-article approval. Due to the near net shape of the casting-- including total net shape of the card guide configuration-- significant weight reduction, a major consideration, was attainable.

Large Titanium Investment Casting Replaces
88 Smaller Aircraft Engine Components

What was once 88 smaller stainless steel (17-4PH) castings, machined and welded together, is now produced as a single 52" (132cm) titanium investment casting with improved strength and dimensional control, and also substantially reduced weight (about 55% the weight of steel with comparable strength).

The part is a fan frame hub used in General Electric's CF-6-80C Engine for A310, A300, Boeing 747, 767, and MD-11 aircraft. It supports the front (fan section) of the engine and ties it to the compressor section .

The technology gained by the production of this part had led to the design and manufacture of many such castings, which means time and money savings for aircraft engine manufacturers. When this casting was first produced, nothing of this size and complexity had ever before been attempted in titanium. The Investment Caster later incorporated more than 380 detailed changes.

Investment Casting Design Cuts Weight Almost in Half, Reduces Assembly, Eliminates Other Metalworking Processes

Investment casting was responsible for cutting the weight of the aluminum housing on this hand-held missile launcher almost in half. This was accomplished by design change, which incorporated the use of thinner walls. The ability to produce castings of this size with .04" walls was a breakthrough and represented "state-of-the-art" for large thin-walled boxes.

The main housing was designed to be complete with handles, and also contain the other two castings as an "as cast" fit that normally would have been a machine-to-fit situation. The three-casting assembly reduced the original design in weight from 4.8 pounds to 2.8 pounds, deleted composite and sheet metal parts and eliminated the necessity of multipart internal assembly. Cast in A357T6, the finished housing measures about 15"x 14" by 7½" high.

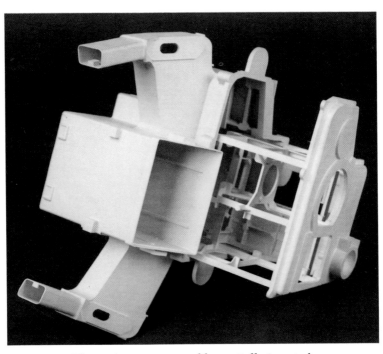

Three-piece cast assembly partially inserted.

Other Typical Investment Casting Components

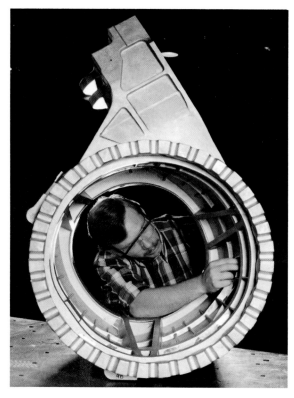

Investment Cast Titanium for Bell Helicopter V-22 Transmission Adapter.

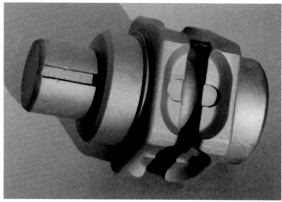

The use of the chain pockets in as-cast configuration make this part ideal for investment casting. To produce this from solid bar, forging or sand casting would require extensive machining to produce a finished part. An investment casting would require only clean up turning of the ends. A tolerance of ±.005 is held on the chain pockets with gaging. Made of 8620 alloy steel.

Other Investment Casting Applications

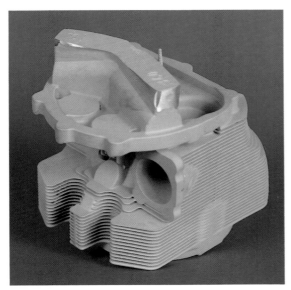

This cylinder head for Lycoming features fins as deep at 2.4" and averaging 1:5". Fins are .060" thick at the tip and .070" at the root. The aluminum (A242) casting weighs 16 pounds and measures 10" long, 7 $\frac{1}{4}$" wide and 8 $\frac{1}{8}$" high.

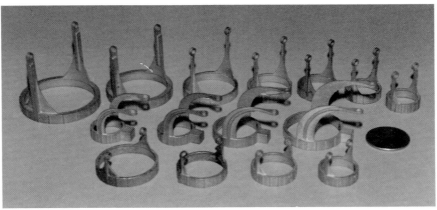

Made of 310 stainless steel, each part has a base ring which includes cast-in-place ridges and part and trademark identification. Arms on some of the most delicate pieces are .060" x .072" x 1.25" long.

Other Investment Casting Applications

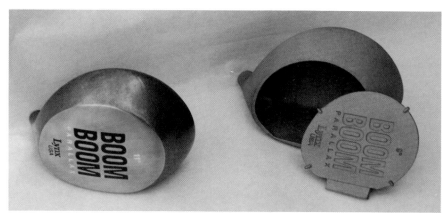

The golf club market is showing remarkable growth. This golf club head is a classic example of the freedom of design allowed by investment casting, which has paved the way for large golf club heads of complex design made of light-weight aerospace metals.

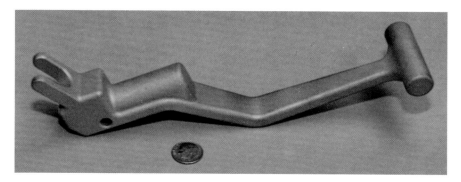

This part, made of CF-8 stainless steel, is used in food service industry. Investment casting eliminates all the grinding and polishing associated with welded joints. It has superior surface finish, which reduces the cost of polishing, which is done by the customer No additional machining required.

Other Investment Casting Applications

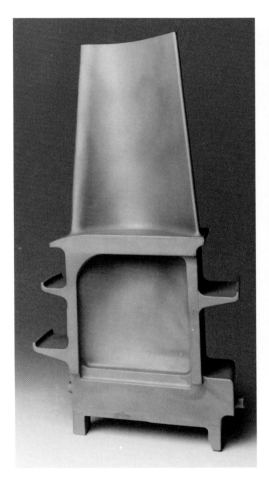 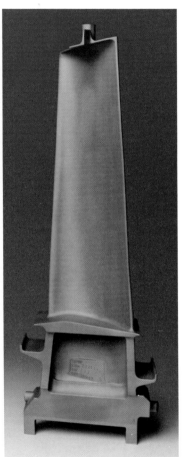

First (at left) and second stage buckets for GE Power Generation's MS9001FA industrial gas turbine measure 17" x 8.5" and 22" x 8" and weight 30 pounds and 28 pounds respectively.

Other Investment Casting Applications

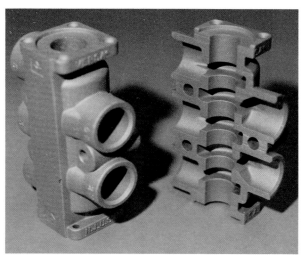

This application is a classic example of a near net shape investment casting. The undercut features (shown in the sectioned casting at right) are difficult and costly to machine. With investment casting, these are easily produced using a soluble wax pattern. The small mounting holes are cast to size including chamfers, ready for tapping. The cast logo, company name and part identification add to marketing appeal.

This fourth-stage vane segment for the MHI/Westinghouse 501f gas turbine measures 24"x23" and weighs 140 pounds.

Other Investment Casting Applications

A cost savings of $25,000 for each part was the result of a machining-to-investment-casting redesign effort. The aluminum casting is a laser chassis for a high-repetition laser incorporated in a night targeting system. Currently made of aluminum (A357-T6), the part is a redesign of a heavily machined or hog-out of wrought aluminum alloy A6061.

This transmission lever neutralizer was previously produced as a ductile iron casting that required a considerable amount of machining. Changing to the investment casting process resulted in costs savings of 50%.

45

Compare These to Your Components

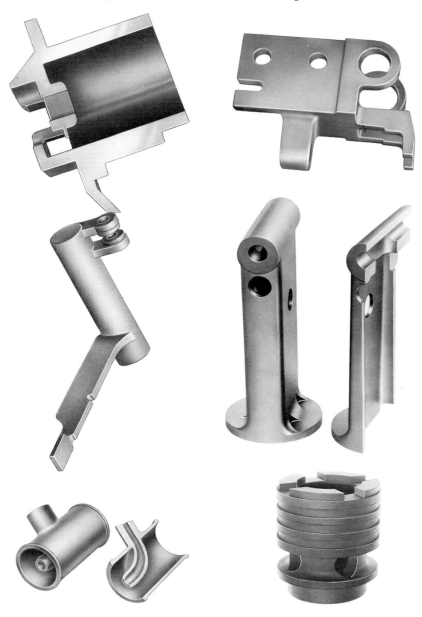

Why not investment cast your components?

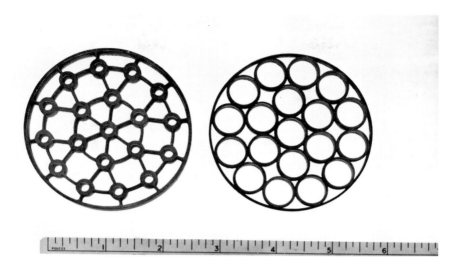

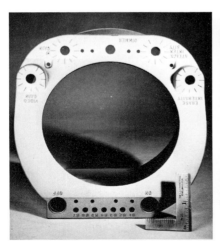

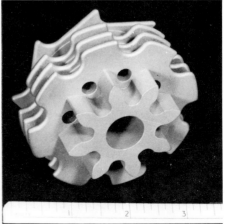

Industries served by the investment casting industry:

      Aerospace: Engines-Airframe
      Automotive
      Bicycle parts
      Defense: Virtually every conceivable area
      Electronic chassis, wave guides etc.
      Farming equipment
      Food processing
      Golf club heads
      Machine tools and accessories
      Medical & dental equipment
      Medical implants; orthopaedic appliances of all sorts.
      Nuclear power generating & control equipment
      Metal working equipment
      Optical equipment
      Petrochemical plant and processing equipment
      Pumps and compressors
      Small arms of all descriptions
      Textile machinery
      Valves of all designs and sizes

This is not an exhaustive list of applications; it is offered so as to stress the myriad of industries that rely upon investment castings from turbine blades operating above their melting points, to golf club heads promising remarkably improved handicap ratings.

Chapter 5
Investment Casting Quality Evaluation, Inspection and Control

Buyers and Investment Casters must agree on the meaning of quality. Each factor affecting quality must be pinned down by specifications and by the purchasing contract which defines the types of casting inspection required and the criteria for acceptance or rejection of the castings. This chapter presents a sound basis for setting up terms of an agreement between producers and customers. Quality may be insured by statistical process control and process capability analysis on the part of the foundry, or by acceptance sampling on the part of the buyer.

Purchase specifications require agreement between the producers and customers. This agreement is generally based on published specifications such as those of the American Society for Testing Materials. Specifications for investment castings should include: primary requirements which apply to all castings and supplementary requirements which are agreed to by the producer and the customer. The primary requirements generally include: heat treatment, alloy composition, workmanship, finish, appearance, quality assurance, weld repair, inspection, certification and product marking. The supplementary requirements may include: dimensional requirements; residual elements; mechanical properties such as tensile strength, hardness or impact toughness; internal soundness determined by x-ray radiography or ultrasonic testing; die penetrant or magnetic particle inspection for casting defects; or microstructural requirements such as grain size.

Definition of Quality

The word "Quality" implies desirability of the product and conformance to particular characteristics or combinations of characteristics which provide suitability for a particular purpose. Suitability includes the degree of conformance to specifications or standards.

Quality is relative and not absolute. Good quality for a particular product when used for one purpose may be quite inadequate if the same product is used for a different purpose. Quality is, thus, relative to the specifications, and to the expected service history for the intended purpose as predicted by reliability indices, warranty periods, etc. Specifications define the standards of quality required under different conditions and state the quality limits in terms of measured

49

variables (length, density, etc.) or attributes such as good surface quality, etc. Specifications and standards simplify decisions on quality and specify the methods of measurement. Specifications are generally set by ballot in balanced committees of specification writing organizations. Such committees are balanced between producers, users and other interested parties. Thus, standards are based on experience with the kind of product concerned.

Rigid quality control standards generally imply a combination of high quality, low quantity and high cost; on the other hand, more flexible controls are apt to result in lower quality standards, greater quantities and lower costs - unless these flexible controls result in more waste or scrap. The decision on specification limits involves balancing losses in sales against the extra cost of "tightening up" the process or providing improved manufacturing facilities.

For a "premium quality" casting the reliability and casting integrity are guaranteed by the foundry. Premium quality connotes not only the designer's requirement for better internal soundness of the casting but also high integrity of the casting; that is, reliability of properties in designated areas of every casting is guaranteed with confidence by the foundry. Quality levels for critical components such as those for aerospace, medical or electronic applications are increasingly being expressed by the "Sigma Level" which expresses the ratio of the specification breadth to the natural process tolerance of the process (mean $\pm$ 3 standard deviations).

Premium quality castings are more costly than ordinary commercial castings. They should be used in highly stressed components for service under severe conditions or substituted for parts fabricated from wrought alloys, where the highest strength-to-weight ratio is essential.

Casting quality is guaranteed by adequate nondestructive testing. Standards have been established for x-ray radiography, magnetic-particle and/or fluorescent penetrant testing. Although these tests are not always sensitive enough to reveal all microshrinkage and inclusions which are harmful to strength, they are essential for detecting any casting not manufactured strictly according to procedure.

Quality Control Procedures

Measures of quality associated with the investment casting process involve processes carried out by the investment foundry, the customer or both. These quality control processes may include: acceptance testing and inspection by the for specific properties contained in the specification; statistical process control (SPC) of measured variables or for attributes; evaluation of process capability using indexes (C_p, C_{pk}, Sigma Level, and ppm defective) which measure process centering and breadth with respect to specification limits; and certification of the quality program according to standards such as the international ISO-9000 standards or the QS-9000 standards for the automotive industry.

a.   Acceptance Sampling:

Acceptance sampling procedures for attributes are detailed in ANSI/ ASQC Z1.4 which replaces MIL STD 105E, and acceptance sampling procedures for variables are specified in ANSI/ASQC Z1.9 which is the civilian replacement for MIL STD 414. Acceptance sampling procedures for attributes and variables along with advanced acceptance sampling techniques are described by Montgomery (1).

b.   Statistical Process Control (SPC):

Statistical process control techniques are used by manufacturers to monitor the quality of items being produced and to warn of process or materials changes which could produce items outside the specification limits. When difficulties are encountered the process is stopped before a significant quantity of bad product is produced.

All processes are subject to small chance variations of various types, and a process is considered to be in control as long as only chance variations are present and the magnitude of the chance variations do not exceed the specification limits.

Specifications and Standards

Specifications are provided by a number of national and international organizations. The national specifications which apply to investment castings may include: military specifications, government specifications, specifications provided by the American National Standards Institute (ANSI) which provides international representation, and specifications provided by professional organizations such as the American Society for Testing Materials (ASTM), the American Society of Mechanical Engineers (ASME), the Alloy Casting Institute (ACI) and other specification writing bodies; and various corporate specifications. The QS-9000 Standards are recent standards set up by Chrysler, Ford and General Motors through the Automotive Industry Action Group (AIAG). The QS-9000 standards cover all aspects of quality associated with automotive products. The international specifications include the International Standards Organization (ISO), the European Standards Organization, and other standards organizations.

The purchase specification is developed by either the maker or the user, who specifies intended conditions of a casting throughout its manufacture or upon its completion. The purchase specification must include not only what is possible, desirable and necessary, but also what is practical, either now or in the future. In the past decade the Uniform Commercial Code (UCC) has been adopted by all 50 states and by the federal government. The UCC governs the rights and responsibilities available under contract law, and it can form the basis for legal action if some provisions of the purchase contract are not followed.

The specification should reflect the intention of its originator, clearly communicating the intentions of either the supplier or the user. A specification written from the supplier's viewpoint may differ greatly from the user's description of the same component. Clearly then, there must be good communications between the customer and the provider.

a.   Specifications

Specifications for investment castings provide information on standard requirements which should be part of the purchase contract for any investment casting and supplemental requirements which can be included by an agreement between the producer and customer. ASTM A732/A732M is a specification for carbon and low alloy steel investment castings and two grades of cobalt-chromium alloy investment castings. ASTM F75 is a specification for cobalt-chromium investment castings used for biomedical implants such as hip or knee prostheses.

ASTM A732 is a good example of a specification for investment castings. Section 1 provides the scope of the specification and Section 2 provides a list of the referenced ASTM documents such as specifications for procedures such as welding, and specifications for testing such as mechanical testing or dye penetrant testing for casting defects. Section 3 provides a list of the information that should be included with ordering castings. This includes information such as the pattern number, alloy designation, the quantity of castings to be ordered and the primary and supplementary requirements.

Sections 4 through 12 provide the standard requirements for investment castings. These include: heat treatment, chemical composition, workmanship and appearance, quality assurance, weld repair, inspection, rejection and rehearing, certification and product marking. Section 4 provides heat treating requirements for carbon and low alloy steel investment castings and defines casting grades based on: heat treatment, tensile and yield strength, and ductility as measured by tensile elongation. The heat treated conditions include: annealed, quenched and tempered, and normalized and tempered. The tensile strengths range from 60 ksi (414 MPa) for Grade 1A in the annealed condition to 190 ksi (1310 MPa) for Grade 12Q in the quenched and tempered condition. Section 5 provides specifications on chemical composition for investment cast carbon and low alloy steels and cobalt-chromium alloys. Procedures are specified for heat analysis, product-check-verification analysis and referee analysis. The alloy specifications include several grades of carbon and low alloy steels along, with two grades of cobalt-chromium alloys.

The supplementary requirements include a number of quality requirements that are designed to insure premium quality castings. These include: limits on residual elements; tension testing at room temperature and elevated temperatures; magnetic particle, liquid penetrant and radiographic inspection; stress rupture test requirements, etc. The purchase contract should call for supplementary requirements where they are needed.

Analysis of the Alloy

Service characteristics of investment casting - aside from shape, finish and dimensions - are determined by the alloy composition and the metallurgical structure of the part. Composition and microstructure control strength, toughness, modulus, corrosion resistance, wear resistance, damping capacity, fatigue life, magnetic and electrical properties and density. Selection of the proper alloy and microstructure is a requirement for ultimate performance of the cast part. Although good design can considerably improve serviceability of the part, the best design and good foundry practice cannot compensate for deficiencies of an alloy which has poor castability or inadequate mechanical properties.

Chapter 6 gives in tabular form the chemical compositions, physical properties, castability and selected industrial applications of investment casting alloy compositions provided by several alloy specification writing bodies. American alloy specification writing bodies include: Federal Specifications, Military Specifications, the Alloy Casting Institute (ACI), the American Iron and Steel Institute (AISI), Society of Automotive Engineers (SAE), American Petroleum Institute (API), American Society for Testing Materials (ASTM) and American Society of Mechanical Engineers (ASME) among others. These alloys are cross listed with foreign alloy specifications for similar compositions. The emphasis on international standards during the last decade has placed a premium on the development of alloy compositions and standards which meet requirements on a worldwide basis.

There are several specifications in common use. Often the specifications for a given alloy will differ in chemical and mechanical properties. In addition, definitions concerning master heats qualification, size limitations if any, and certification procedures will vary even within the controlling source (e.g. ASTM, AMS). IT IS ESSENTIAL FOR THE USER AND INVESTMENT CASTER TO BE AWARE OF, AND AGREE UPON, THE PRECISE REQUIREMENTS OF THE SPECIFICATION REQUIRED.

Mechanical Properties

Castings and test bars must be mechanically tested to certify that mechanical property specifications have been met. The mechanical property tests include: tensile tests, hardness tests, impact toughness tests, stress rupture tests, fatigue tests, or other tests required by specifications. Test specimens may be attached to the castings or cast separately. Mechanical test requirements are covered in ASTM A 732/A732M-94 for investment castings and in ASTM A370-95 for steel products in general. In recent years the responsibility for writing and maintaining specifications has been passed from the branches of the federal government (such as Department of Defense) to various professional and trade organizations which are associated with production or with product utilization. These organizations prepare specifications for the American National Standards Institute (ANSI) which represents the United States with the International Standards Organization (ISO).

The tensile test is most often required in specifications, since it provides a means of measuring strength and ductility of metal under standard and comparable conditions. Cast test bars indicate only the quality of the metal from which the casting is made. They do not give actual properties of the casting; neither are they a quantitative measure of casting quality. They are not truly representative of the final casting. The chief value in tensile testing lies in assuring consistency of metal properties from heat to heat or from one casting lot to the next.

It is important that test specimen design be uniform throughout the industry. Figure 5:1 shows the design of the standard Investment Casting Institute test bar which is also the test bar used for the ASTM A732 specification for investment castings.

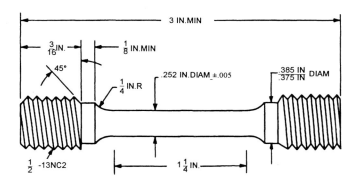

Figure 5:1 Design and dimensions of the Investment Casting Institute test bar

Increasing the .375"-.385" shoulder diameter to a maximum of .500" is permissible. Although the test bars are for evaluation of metal only, casting conditions should be similar to those used in production. Threads may be machined, but machining must be done in a manner to minimize stress over the gage length. The entire bar may be machined and it is permissible to decrease the diameter a maximum of .003" at the center to facilitate midpoint fracture.

Hardness

Hardness measures the resistance of metals to plastic deformation by indentation. Table 5-1 below lists the various hardness tests, the indentor type and the method of measurement.

Carbon steels and low alloy steels show a good correlation between hardness and ultimate tensile strength. Hardness is a good indication of consistency and when correlated with the composition and heat treatment of cast ferrous and nonferrous alloys it is a handy measure of the control and acceptability of castings. The low cost and simplicity of hardness tests leads to their use for a number of quality control applications.

Table 5-1: Hardness Testing

Test	Indentor	Measurement
Brinell Hardness ASTM E10-93	1/4" steel or tungsten carbide ball	Indentation Diameter
Rockwell Hardness	1/16" or 1/8" steel ball or conical diamond indentor	Penetration Depth
Diamond Pyramid	diamond pyramid	Ave. of indentation diagonals
Vickers Hardness	diamond pyramid	Ave. of indentation diagonals
Knoop Hardness	elongated diamond pyramid	Ave. of indentation diagonals

Meeting Specified Dimensional Tolerances

Investment casting is distinguished from other casting processes by its ability to hold close tolerances for external and internal dimensions. This ability has enhanced the value per pound of investment castings, and has led to the use of the investment casting process for critical applications such as airfoils, biomedical implants, or weapons components.

Chapter 3 provides information on the tolerances recommended and accepted by the Investment Casting Institute. More rigorous tolerances are possible, but closer tolerances will depend on the specific part, the investment casting producer and the justification for increasing cost. If you have a problem which requires tolerances closer than those which are generally accepted as standard, you should contact your investment casting supplier to discuss the details of the requirement.

Inspection of Castings

Reference should be made to Chapter 6 which describes the voluntary Quality Control Standards developed by the Investment Casting Institute defining the metal quality of normal commercial investment castings.

ASTM A732, the inspection of investment castings is a general quality control requirement for most purchase contracts. Inspection techniques include: visual inspection, magnetic particle inspection, liquid penetrant inspection, radiographic inspection and ultrasonic inspection.

a.   Visual Inspection:

Certain types of defects are obvious upon visual examination of the casting: cracked castings, tears, slag adhesions, blowholes, metal penetration, scabs, shifts, non-fill, cracked molds or cores and similar defects can be identified visually.

(1) Inspection of as-cast or cast and machined surfaces for adherence to visual acceptance standards shall be by the naked eye in sunlight or normal white light without benefit of magnification unless otherwise agreed upon between the customer and the foundry. Only in very special circumstances, however, will visual inspections be performed with more than five magnifications. Treatments such as acid etching to remove surface metal prior to inspection will not be done unless agreed upon between the customer and the foundry.

(2) Acceptance for visual standards will be based upon surface appearance only, exclusive of rms or AA values and without consideration of magnetic particle, fluorescent penetrant, radiographic, or other nondestructive testing requirements for findings.

(3) Acceptable positive surface irregularities, as noted below, must fall within the tolerance range of the surfaces involved unless otherwise agreed upon between the customer and the foundry. Acceptable negative surface irregularities will not be used as a reference for dimensional measurements of the surfaces involved.

Surface irregularities should be classified as follows:

(1) Positive surface irregularities on castings may result from excess metal caused by metal penetration into the ceramic mold material, or from air or liquid entrapment next to the pattern during the investing operation.

(2) Negative surface irregularities on castings may be caused by slag or oxide inclusions, gas, shrinkage, oxidation pitting or cracks.

(3) Positive surface defects, generally referred to as excess metal, are not potential stress raisers and acceptance evaluation usually is based upon aesthetic values, provided that tolerance requirements have been met. Negative surface irregularities may be either propagating or non-propagating. Propagating indications include visual surface cracks or cold shuts, while non-propagating

indications include oxidation pitting, oxide or slag inclusions and shallow linear depressions.

Evaluation of visual standards is an area subject to differences in opinion because it depends almost completely upon human judgement. Extremely close cooperation and agreement on mutual problems is essential between the customer and the foundry. Standards should be based upon the widest acceptability possible.

The Investment Caster can produce parts to almost any quality level required by the customer. A standard should always be set up between the customer and the Investment Caster. This standard may be fairly rigid or quite loose, depending on the customer's need. A set of quality standards has usually been accumulated by the Investment Caster from his experience. Drawing upon this experience, the customer and Investment Caster usually agree on inspection procedures and limits to be set.

b.   Magnetic Particle Inspection:

Magnetic particle inspection is a nondestructive method for detecting cracks, seams, inclusions, segregation, porosity and similar discontinuities in magnetic materials. It is not applicable to non-magnetic materials. This method will detect surface discontinuities which are too fine to be seen by the naked eye, defects which lie slightly below the surface and - with special equipment - even deep-seated discontinuities.

Not all discontinuities in metal are detrimental to its efficient service. The inspector must be able to interpret magnetic particle indications and decide which discontinuities are to be regarded as defects. Since one can expect a wide variation in evaluation of the results, the following points should be agreed upon when inspection is being considered:

1.   What techniques to use (specified in detail);
2.   What types of discontinuities shall be rejected;
3.   What types of discontinuities may be accepted;
4.   The definition of reworking and subsequent retesting which may be permissible.

In magnetic particle inspection a magnetic field is induced in the piece to be inspected and the piece is covered by finely divided magnetic particles. The magnetic field escapes from the work piece at discontinuities, and the magnetic particles align themselves with magnetic poles produced at the edges of the discontinuity to form a pattern which outlines the shape of the discontinuity. Information on magnetic particle inspection is provided by the following specifications:

ASTM E 1444-94a	Practice for Magnetic Particle Inspection
ASTM E 709-95	Guide for Magnetic Particle Inspection
ASTM E 125-63	Reference Photographs for Magnetic Particle Indications on Ferrous Castings (revised 1993)

c.   Liquid Penetrant Inspection

Liquid penetrant inspection is a sensitive, nondestructive method for detecting minute discontinuities (cracks, porosity, holes or surface seams) in non-magnetic materials where magnetic particle inspection cannot be used. A liquid with high penetrating qualities is applied to the part surface and is drawn into extremely small surface openings by capillary action. The excess liquid is removed, and a developer such as talcum powder is applied to the surface. The penetrant trapped in discontinuities flows back up to color the developer and provide an indication.

In dye penetrant testing the liquid penetrant is colored with a visible dye, and the surface is inspected under normal white light. In fluorescent penetrant testing the liquid penetrant contains a fluorescent dye and the surface is inspected under fluorescent light. Information on liquid penetrant testing is contained in the following specifications:

ASTM E 165-95	Liquid Penetrant Testing
ASTM E 1208-94	Fluorescent Liquid Penetrant Examination Using the Lipophilic Post-Emulsification Process
ASTM E 1210-94	Fluorescent Liquid Penetrant Examination Using the Hydrophilic Post-Emulsification Process
ASTM E 1219-94	Fluorescent Liquid Penetrant Examination Using the Solvent-Removable Process
ASTM E 1209-94	Fluorescent Liquid Penetrant Examination Using the Water-Washable Process
ASTM E 1220-92	Visible Liquid Penetrant Examination Using the Solvent-Removable Process
ASTM E 1418-92	Visible Penetrant Examination Using the Water-Washable Process
ASTM E 1135-86	Comparing the Brightness of Fluorescent Penetrants
ASTM E 1417-95a	Practices for Liquid Penetrant Examination
ASTM E 433-86	Reference Photographs for Liquid Penetrant Inspection (revised in 1992)

d.   Standards for Magnetic Particle and Liquid Penetrant Inspection

Magnetic particle and dye penetrant inspections should conform to the following:

(1) Indications of themselves shall not be cause for rejection unless so specified by the customer.

(2) At the time of inspection the nature of the discontinuities shall be determined and placed on one or more of the following categories:

a. Propagating indications
b. Nonpropagating indications
c. Dispersed indications

(3) Propagating indications are those which conceivably could increase in length or width under stress, resulting in failure of the part. In this category would fall such indications as cracks, hot tears, segregation, pits, holes, or other imperfections so aligned as to be propagating in nature.

(4) Nonpropagating indications are discrete, usually rounded indications. This category would include gas holes, holes resulting from air entrapment, dross holes, inclusions, pits and other surface imperfections.

(5) Dispersed indications are those which are nonpropagating in nature but are distributed over a larger area of the casting. In this category would fall sponge and microshrinkage (generally found only in aluminum and magnesium castings), gas porosity and other surface indications.

e.   Ultrasonic Inspection

Advances in ultrasonic testing have led to the widespread application of ultrasonic scans to detect shrinkage porosity, cracks and other defects deep within castings. In the simplest type of ultrasonic inspection ultrasonic waves are initiated at the surface by a high frequency transducer, they reflect from defects or from the rear surface of the casting and are picked up by the transducer. A plot of the amplified reflected waves on a CRT shows the presence of defects within the casting. More sophisticated ultrasonic testing techniques involve immersion of the casting in a water bath enhancing the pickup of the ultrasonic waves to provide a two dimensional map of the defects present. These techniques are applied to a wide variety of castings, ingots and forgings. Ultrasonic testing is covered in ASTM Specifications:

ASTM E 114-91	Practice for Ultrasonic Pulse-Echo Straight-Beam Examination by the Contact Method
ASTM E 214-68	Practice for Immersed Ultrasonic Examination by the Reflection Method Using Pulsed Longitudinal Waves
ASTM E 587-94	Practice for Ultrasonic Angle-Beam Examination by the Contact Method
ASTM E 1001-90	Detection and Evaluation of Discontinuities by the Immersed Pulse-Echo Ultrasonic Method Using Longitudinal Waves

f.   Radiographic Inspection

Background:

In 1958 a committee under the auspices of the ASTM began to formulate a group of radiographs for referencing thin-walled castings. This work culminated in ASTM Specification E 192-62T, Reference Radiographs of Investment Steel Castings for Aerospace Applications. This specification was adopted by the military, and MIL-Std-2175 makes use of this specification for thin-walled steel castings. Reference radiographs were also issued by the Navy under specification NAVWEPS 00-PK-500. In 1966 the ASTM adopted E 192 as a permanent specification, and it is now ASTM E 192-91 with the latest revision in 1991. Although both the ASTM specification and MIL-Std-2175 are referenced most often for aerospace applications, their use is also prevalent for commercial investment castings.

There are significant benefits in having a set of reference radiographs, but their use can be dangerous unless those persons who set the standards are familiar with the specific requirements of individual parts and the effect of discontinuities on their function. There is a tendency to overgrade or over classify a casting or an area of a casting. The military specification - cognizant of this tendency - issued a warning in the last paragraph of MIL-Std-2175:

"Producibility Warning. Caution should be exercised in specifying the grade of maximum permissible radiographic discontinuity level to be met by a casting.

"Casting design coupled with foundry practice can make too severe soundness requirements impractical for a manufacturer to satisfy. The class assigned to the casting should represent a realistic value for the functional requirement, i.e., do not assign a class 1 casting for a class 4 function."

Casting Class:
MIL-Std-2175 divides castings into four classes, 1, 2, 3 and 4 with each class subdivided twice for a total of eight classes: 1A, 1B, 2A, and 2B. The classification defines the functional requirements for a casting.

Class 1- Castings whose failure would result in the loss of a missile, an aircraft - or any other type of vehicle - or would cause significant danger to operation personnel.

Class 2 - Castings whose failure would result in the loss of a major component of a missile, aircraft or other vehicle - or would cause significant danger to operation personnel.

Class 3 - Castings other than Class 1 castings having a margin of safety of 200% or less.

Class 4 - Castings other than Class 1 castings having a margin of safety greater than 200%.

Class 1 castings shall be completely examined, and Class 2 and 3 castings shall be radiographed according to a sampling plan. In practice, it has been found that Class 2 castings are radiographed 100%. A Class 4 casting does not require radiographic examination.

Interpreting Discontinuities:

Radiographic discontinuities are interpreted using ASTM E 192-91 which provides reference radiographs to illustrate the types and degrees of discontinuities that occur in investment castings. The radiographs are applicable to cast sections up to one inch (25.4 mm) and they cover graded discontinuities with eight degrees of increasing severity and twelve ungraded discontinuities.

Chapter 6

Alloy Selection for Investment Casting
Ferrous, Nonferrous, Airmelt, Vacuum

The data in the following tables relating to air melted alloys is derived from the Institute's publication *'Basic Metal Quality Standards for Investment Castings'*. The tables contain preferred chemistries and mechanical properties obtained from investment cast to size 1/4" diameter test bars.

In this chapter some general information is given on commonly cast alloys.

Alloy Selection

Several factors must be considered when selecting an alloy for investment casting. Often the designer has a choice between alloys, any one of which would perform satisfactorily in the intended application. Five areas should be examined before the final alloy choice is made. Each of these areas will be dealt with briefly. They are as follows:

A. Environment and intended use of the part.
B. Mechanical property requirements.
C. Foundry characteristics of the alloy.
D. Fabricating characteristics.
E. Cost factors.

Investment castings are used in such diverse environments as air, fresh and sea water, soil, hot gases, corrosive chemicals and space—at temperatures ranging from cryogenic to 2200° F. Exposure to a particular environment may be continuous, cyclic or intermittent, and the intended life of the part may be measured in years or in seconds. The casting may be required to function under pressures of microns or upwards of 200,000 pounds per square inch. The use of a more expensive vacuum-melted and vacuum-cast alloy may be indicated for some of the most severe environmental conditions.

Closely related to environment and use are mechanical property requirements. The laboratory conditions under which many mechanical property tests are performed cannot duplicate actual service conditions; however, such test data are useful in alloy selection so long as the limitations are kept in mind. Evaluation of hardness, short time tensile properties, stress rupture and creep properties, impact strength, wear and corrosion resistance against the type of environment to be encountered will further narrow selection of an alloy.

Foundry characteristics become increasingly important with increasing size and/or complexity of the casting to be made. Castings with thin sections require an alloy with good fluidity and intricate castings will require an alloy of high resistance to hot tearing in addition to good fluidity. Fluidity is the ability of a liquid metal to run into and fill a mold cavity. Resistance to hot tearing is the ability of a metal to resist tearing or cracking during the last stage of solidification. It is related to solidification shrinkage and the tendency to hot tearing can be partly compensated for by designing the casting so as to permit unhindered shrinkage. The natural shrinkage of a metal as it passes from the liquid to the solid state can result in unsound castings. This solidification shrinkage must be allowed for in the design of a gate and riser system so as the casting solidifies and contracts, an adequate supply of molten metal is available in the mold to continually fill the casting cavity and prevent the formation of shrinkage voids or unsound areas. A large gating system will result in correspondingly lower foundry yield.

Castability

The foundry characteristics of an alloy are combined in the general term "castability." In addition to fluidity and tear resistance mentioned above, castability includes reaction of the molten metal with its environment (atmosphere, crucible, mold) and its freezing characteristics. How these factors are handled will determine the degree of production success attained. Castability characteristics are obviously not the same as those required for good service performance. Since there are no castability tests which enjoy the acceptability of mechanical and physical property tests, ratings are somewhat arbitrary—determined by evaluation against an alloy with which the foundry industry has had little or no difficulty.

Chart 1 shows the castability ratings for a number of alloys cast by the investment casting industry.* Evaluation is based on a relatively simple casting configuration and upon comparisons with three alloys having excellent foundry characteristics and assigned a castability rating of 100. These are: 302 stainless steel (ferrous), 20C beryllium copper (nonferrous) and aluminum alloy 356. The fluidity, shrinkage and hot tear ratings of each alloy listed are classified as 1-best, 2-good and 3-poor.

Fabricating characteristics, in addition to casting, include such varied operations as heat treatment, machining and grinding, welding and straightening.

Most investment cast alloys require some type of heat treatment before use. The heat treatment may range from a simple anneal to lengthy procedures involving sub-zero temperatures and long furnace times at high temperatures to develop required or desired properties.

Machinability is a composite characteristic and refers to the relative ease of metal removal. With the difficult-to-machine alloys, grinding is probably the most widely used metal removal technique.

Welding of investment castings—in fabricating larger assemblies or repairing defects—can be performed successfully using established processes. Some alloys are designed to be more readily welded than others. Which process and the type of pre- or post-welding heat treatments are important considerations in producing good weldments. Straightening may be necessary following heat treating or welding.

This chart is based on the results of a survey of investment casting

CHART 1
CASTABILITY RATING OF INVESTMENT CASTING ALLOYS

Alloy	Castability Rating	Fluidity	Shrinkage	Resistance to Hot Tearing	Alloy	Castability Rating	Fluidity	Shrinkage	Resistance to Hot Tearing
Silicon Irons (Electrical alloys of pure iron and silicon)					**400 Series Stainless** (A.I.S.I. designations)				
0.5% Si.	75	3	3	2	405	90	2	3	2
1.2% Si.	80	3	3	2	410	95	1	3	2
1.5% Si.	80	3	3	2	416	85	1	3	2
1.8% Si.	75	3	3	2	420	90	1	3	2
2.5% Si.	70	3	2	2	430	90	1	3	2
					430 F	90	1	3	2
Carbon & Sulphur Steels (A.I.S.I. designations)					431	90	1	3	2
					440 A	85	1	3	2
1015	80	3	3	3	440 C	85	1	3	2
1018	80	3	3	3	440 F	85	1	3	3
1020	80	3	3	3	AMS 5355				
1025	85	3	3	2	(Armco 17-4 PH)	85	1	3	2
1030	85	3	2	2	AMS 5354	85	1	3	2
1035	85	3	2	2					
1040	85	2	2	2	**300 Series Stainless** (A.I.S.I. designations)				
1045	85	2	2	2	302	100	1	1	1
1050	85	2	2	2	303	95	1	1	2
1060	85	2	1	2	304	100	1	1	1
1117	75	3	3	3	310	90	1	1	1
1140	80	2	2	3	312	90	1	1	1
					316	100	1	1	1
Low Alloy Steel (A.I.S.I. designations)					347	95	1	1	1
					CF-8M (ACI)	100	1	1	1
2345	90	2	2	2	CN-7M (ACI)	95	1	1	1
3120	85	2	3	2					
4130	90	2	2	2	**High Nickel Alloys**				
4140	90	2	2	2	Monel (QQ-N-288-A)	85	1	2	2
4150	90	2	2	2	Monel, R.H.	75	1	2	3
4340	90	2	2	2	Monel, S. (QQ-N-288-C)	75	1	2	3
4615	85	2	3	2	Inconel (AMS-5665)	85	1	2	2
4620	85	2	3	2	47-50 (47% Ni-50% Fe)	80	3	2	2
4640	90	2	2	2	Invar	75	1	2	2
5130	85	2	3	2					
6150	90	2	2	2	**Cobalt Alloys**				
8620	85	2	3	2	Cobalt J	80	1	1	3
8630	85	2	3	2	Cobalt 3	80	1	1	3
8640	90	2	2	2	Cobalt 6 (AMS-5387)	80	1	1	2
8645	90	2	2	2	Cobalt 19	85	1	1	3
8730	85	2	3	2	Cobalt 21 (AMS-5385C)	90	1	1	2
8740	90	2	3	2	Cobalt 31 (AMS-5382B)	90	1	1	2
52100	80	1	2	2	Cobalt 93	70	1	1	3
Nitralloy	75	2	3	2	N-155 (AMS-5531)	80	1	1	3

Castability ratings are based on a casting of relatively simple configuration and upon comparisons with three alloys having excellent foundry characteristics and assigned castability rating of 100. These are 302 stainless steel (ferrous), 20 C beryllium copper (nonferrous) and aluminum alloy 356. The fluidity, shrinkage and hot tear ratings of each alloy are based on 1--best, 2--good, and 3--poor.

CHART 1 (Cont'd) CASTABILITY RATING OF INVESTMENT CASTING ALLOYS										
Alloy	Castability Rating	Fluidity	Shrinkage	Resistance to Hot Tearing	Alloy	Castability Rating	Fluidity	Shrinkage	Resistance to Hot Tearing
Tool Steels					**Copper Base**				
(A.I.S.I. designations)					Al. Bronze Gr. C	80	1	3	1
A-2	85	2	2	2	Al. Bronze Gr. D	80	1	3	1
A-6	80	2	2	2	88-10-2 (G Br. &				
D-2	85	3	2	2	Gun Metal)	85	1	3	1
D-3	85	3	2	2	Mn Bronze	80	2	3	1
D-6	80	3	2	2	Hi Tensile Mn Bronze	80	2	3	1
D-7 (BR-4)	80	3	2	3	Naval Brass				
BR-4 FM	80	3	2	2	(Yellow Bronze)	85	2	2	1
F-2	75	3	2	2	Navy "M"	85	2	2	1
H-13	85	2	2	2	Navy "G"	85	2	2	1
L-6	80	2	2	2	Phosphor Bronze				
M-2	80	2	2	3	(SAE 65)	85	2	2	1
M-4	75	2	2	3	85-5-5-5 (Red Bronze)	90	1	1	1
O-1	80	2	1	2	Silicon Brass	100	1	1	1
O-2	80	2	1	2	Be Cu 10C	90	1	1	1
O-7	80	2	1	2	Be Cu 20C	100	1	1	1
S-1	90	2	2	2	Be Cu 275C	90	1	1	1
S-2	90	2	2	2					
S-4	90	2	2	2					
S-5	90	2	2	2					
T-1	80	2	1	3					
Aluminum									
13	85	2	2	2					
40E	75	3	3	3					
43	90	2	2	2					
356-A356	100	1	1	1					
355-C355	95	1	1	1					
B-195	85	1	2	2					

Cost is a factor in all of the foregoing. Frequently, material cost is of minor importance compared with the costs of the environmental and manufacturing considerations. Sometimes, selecting a costly alloy on a pound-for-pound basis will result in a less costly finished part: the more costly alloy may be easier to cast, hold more consistent dimensions, resist hot tearing and have a better as-cast finish. Again, an alloy which appears more costly at first glance may pay dividends in longer part service life or higher strength-to-weight ratios for aircraft and space application. In every case, the purchaser and investment casting vendor should agree upon the best alloy prior to manufacture.

Investment Casting Alloys

The alloys most commonly investment cast and their ASTM specifications are listed in Tables 1-17.

There are nine independent major materials specifications writing groups in the United States. The trend is for those groups to conform to the American Society of Testing Materials specifications in the hope of reducing the number of specifications, thus simplifying the jobs of users and producers of investment castings. The Institute endorses this for non-aerospace castings.

Wherever possible, the chemical range of an existing selected specification is used in order to avoid a variety of chemical ranges for the same alloy.

Data in charts and tables is to be used as a guide only in the selection of alloys. This information, modified and supplemented by experience, should be the basis for a final choice. In any unusual situation, the advice of an experienced producer should be sought.

Alloy Names

Alloys are listed by their most common identifying name or number designation and are grouped by the basic element contained. **Trade names are used only for the purpose of type identification.**

All chemical specifications are for cast alloys; where reference is made to wrought specifications it is only to indicate similarity.

Composition Range

The chemical compositions given are not specifications. Alloys of the same type may have different ranges for particular elements depending upon processing and end product requirements.

Complete chemistries are not given. Only the important alloying elements have been listed. Not included are those residual or tramp elements which are permitted only up to specified maximum values. The major or base element is indicated in the column headings for each group of alloys.

Low silicon values are typical of many alloys in wrought form. It is customary for castings to be produced with higher silicon contents—generally up to one percent maximum.

Code numbers

This is the basic key to identification and cross reference as taken from the MIL-HDBK-HIC referred to earlier. The code groups materials of similar composition by number and each number may represent several specifications.

Code numbers are not specifications and cannot be used as such for procurement or any other purpose. Specifications represented by the same code number may or may not have similar mechanical properties or other characteristics and, therefore, cannot be substituted for each other indiscriminately.

Only those code numbers of primary interest for castings (and most representative of the alloys) have been listed.

Specifications

Each Institute alloy chart includes typical mechanical properties for preferred alloys when given the usual heat treatment and indicates certain alloy applications and characteristics.

AMS

The Aerospace Material Specifications are prepared by the Aerospace Materials Division of the Aerospace Council. They are obtainable at low cost from the Society of Automotive Engineers, Inc., 485 Lexington Avenue, New York 10017.

The Aerospace Material Specifications are complete procurement specifications for materials used in the manufacture of aircraft, aircraft engines, propellers, missiles and other aircraft accessories. They include requirements regarding form, quality, test reports, etc., and are the basis for acceptance or rejection of such purchased materials and/or parts.

Chemical or physical composition of materials covered by the AMS are coordinated as far as possible with SAE general standards for similar materials, but where necessary, the limits of acceptable composition may be more restrictive.

The symbol "AMS" preceding the Aerospace Material Specification number is an integral part of the identification and should always be included in referring to individual specifications by number. Revised or amended specifications are indicated by letter suffixes, e.g. AMS-5362D is the fourth revision of AMS-5362.

Letter suffixes indicating revisions or amendments have not been included since it is considered advisable to request the latest revision

ASTM

Specifications of the American Society for Testing and Materials are based on use and quality. They are reprinted annually in book form by the Society. Each ASTM specification consists of a letter indicating the general classification (A denotes ferrous material; B denotes nonferrous material), a serial number and a second number indicating the year of adoption or last revision.

SAE

The Society of Automotive Engineers has a system for classifying steels based on chemical analysis of the steel. It is widely used and accepted; however, the system is for wrought ferrous materials only and, therefore, has definite limitations so far as the Investment Caster is concerned. These compositions must be modified for casting applications.

The first number in an SAE designation indicates the type of steel. The second number generally indicates the approximate percentage of the chief alloying element. In some cases, two digits are needed for this second number. The last two figures indicate the nominal carbon content in "points" or hundredths of one percent. Since these are wrought alloy compositions, the reader is referred to the SAE Handbook for a detailed explanation of the SAE numbering system.

QQ

This is the symbol used to identify Federal specifications for metals. They are used by the various departments of the United States Government. Suffix letters and numbers indicating revisions and amendments respectively to the specifications are not included in this index. Copies of Federal specifications and an index can be purchased from the Superintendent of Documents, U.S. Government Printing Office, Washington, D.C. 20025. At the time of publication Mil specifications are being withdrawn and replaced by alternate specifications.

MIL

This symbol identifies military specifications as issued under the authority of the Standardization Division, Office of the Assistant Secretary of Defense (Supply and Logistics). It should precede the specification number in every case. There are two types; "Coordinated" specifications covering items of common use by the three military departments and "Limited Coordinated" as issued by a single departmental interest. Copies may be obtained from procuring agencies requiring these specifications or as directed by the contracting officer.

ACI

The Alloy Casting Institute is composed of a group of foundries producing high alloy, heat resistant and corrosion resistant castings. It has developed a system of standard nomenclature for these alloys. The chemical composition ranges covered by these ACI designations have been adopted by other groups. An initial letter "C" in these designations indicates alloys generally used to resist corrosive attack at temperatures less than 1200° F. The initial letter "H" indicates an alloy for use generally at temperatures in excess of 1200° F. The second letter indicates the nickel content ranging from "A" (indicating 1% max.) to "X" (indicating 64-68% nickel).

Voluntary Standards for Commercial Investment Castings

I. Scope

This voluntary standard describes the metal quality of commercial grade castings of the more common alloys to be provided by participating Investment Casting Institute members in situations where purchasers of such castings do not provide detailed specifications covering all aspects of metal quality.

II.   PURPOSE

To define a normal level of metal quality to be furnished by participating Investment Casting Institute member companies as a service to purchasers of investment castings who do not cite detailed specifications. Thus, a casting purchased from one participating Investment Casting Institute member should be of the same basic quality level when purchased from any other member.

III.   GENERAL

A.   There are nine independent major materials specifications writing groups in the United States. The trend is for those groups to conform to the American Society of Testing Materials specifications in the hope of reducing the number of specifications thus simplifying the jobs of users and producers of investment castings. The Institute endorses this for non-aerospace castings.

B.   For each alloy this standard, wherever possible, uses the chemical range of an existing selected specification in order to avoid a variety of chemical ranges for the same alloy.

IV.   METALLURGICAL STANDARDS

General
1. This section shows the chemical ranges considered standard by the Institute.
2. When chemical analysis is done, it will be done by spectrometric, x-ray fluorescence, or other approved methods using comparative standards traceable to the National Bureau of Standards.

Aluminum Alloys
1. Table 1 shows the trade names for the aluminum base alloys normally cast and the standard chemical analysis ranges. Table 2 shows some mechanical property ranges these alloys can provide.
2. When heat treatment is required it will be performed in accordance with good commercial practice.
3. Unacceptable flaws may be removed and the metal replaced by a weld deposit using filler of the same alloy prior to final heat treatment The welded area shall meet the quality standards of the base metal.

High Cobalt Alloys
1. **Table 3** lists the trade names and chemical ranges for this group of alloys. **Table 4** shows some mechanical properties they can provide.
2. Unacceptable flaws may be removed and the metal replaced by a weld deposit using the same filler. All welds will be penetrant inspected .
3. Parts not heat treated after weld repair will be given a post weld stress relief unless otherwise approved by the purchaser.

Copper Alloys

1. **Table 5** shows the trade names for the copper base alloys normally cast and the standard chemical analysis ranges. **Table 6** shows some mechanical property ranges these alloys can provide.
2. If required, heat treatment will be done in accordance with good commercial practice using times and temperatures appropriate for the alloy.
3. Unacceptable flaws may be removed and the metal replaced by a weld deposit using the same alloy as filler prior to final heat treatment (if required). The quality of the welded area shall meet the quality standards of the base metal.

High Nickel Alloys

1. **Table 8** gives the trade names and chemical ranges of this group of alloys. **Table 9** shows some of the mechanical properties they can provide.
2. The nickel alloys are readily weldable; unacceptable flaws may be removed and the metal replaced by a weld deposit using filler of the same alloy prior to any final heat treatment.
3. When heat treatment is required it will be done in accordance with good commercial practice.

Iron Base Alloys

1. Heat Treatment and Decarburization
   a. Heat treatment will be done in accordance with good commercial practice. Some, as cast alloys in this group, normally have a degree of decarburization. When surface hardness is required by the application carbon restoration on those alloys should be specified and confirmed by hardness test.
   b. For stainless steels solution annealing is recommended for maximum corrosion resistance

2. Iron, Carbon and Low Alloy Steels
   a. **Table 10** shows the alloy designation and chemical analysis ranges for this group of alloys. **Table 11** shows some of the mechanical properties achievable from these alloys.
   b. Unacceptable flaws may be removed and the metal replaced by a weld deposit using the same alloy as filler for alloys up to 0.4% carbon nominal and the welded areas will meet the quality standards of the base metal. Above 0.4% carbon nominal, filler of the same alloy is not normally suitable and when it is used the purchaser and foundry should agree on the type, extent and location of welding to assure it is consistent with the intended use. All welding will be done prior to final heat treatment.

3. Hardenable Martensitic Stainless Steels
   a. **Table 12** shows the trade names and chemical ranges considered standard for this group of alloys. For convenience, three age hardenable alloys are included in the table. **Table 13** shows some of the mechanical properties that those alloys can achieve.
   b. In all alloys except 440A, 440C, and 440F, unacceptable flaws may be removed and the metal replaced by a weld deposit using the same alloy as filler metal prior to final heat treatment. The welded areas will meet base metal quality standards. Welding of the 440 series of alloys will be as agreed between the purchaser and the foundry, depending on the use and nature of the part .

4. Austenitic Stainless Steels
   a. **Table 14** shows the trade names and chemical range for these steels. The ACI designations (e.g. CF-8) are being more widely used and should eventually replace the use of the 300 series names, which became accepted as wrought alloy specifications. **Table 15** shows some of the mechanical properties of this group of alloys.
   b. All alloys in this group are readily welded; unacceptable flaws may be removed and the metal replaced by a weld deposit using the same alloy as filler prior to final heat treatment. All welds will meet the base metal quality requirements. Limits on post heat treat welding will be as agreed between the foundry and its customer.

5. Tool Steels
   a. There are few casting specifications for this group of steels; their use as investment castings is specialized. Cast tool steels use the wrought, SAE nomenclature except that a letter "C" precedes the name. This is necessary because the chemical specifications for cast tool steel often differ from the wrought counterpart. They will be supplied to meet chemical analysis ranges as shown in **Table 16**. Standard Hardness values only are listed for mechanical properties. **(Table 17)**.
   b. Except for H-11, H-12 and H-13 which may be welded like low alloy steels, unacceptable flaws will be removed and the metal replaced with a weld deposit using filler of the same alloy, on non-working surfaces only. If other filler is used, or parts are to be welded on working surfaces, prior approval will be obtained from the purchaser. All welds will be magnetic particle inspected or penetrant inspected prior to final heat treatment.

V.   QUALITY CONTROL STANDARDS
General
1. The Institute strongly recommends that each purchaser work out detailed requirements for metal quality for each part. These requirements include but are not limited to: appearance, surface finish, surface cleanliness, soundness, non-fill and other special requirements.
2. This section of the standard defines what Institute members consider the metal quality of a normal commercial investment casting. The Institute cannot guarantee the adequacy of these standards for any part since design and intended function are controlled by the customer. If a part requires characteristics other than these described it is the responsibility of the purchaser to generate and specify those requirements prior to requesting a quote for the part.

Chemical Analysis
The foundry will control its melting stock and melting practices to provide castings that meet the analyses specified.

Mechanical Properties
1. Mechanical properties of the heats or castings will not be determined unless specifically required by the purchase order. If not mentioned in the request for the quote and required by purchase order its cost will be extra and borne by the purchaser.
2. Tensile properties and/or hardnesses are the most commonly specified mechanical properties and, **for information only**, Tables 2, 4, 6, 9, 11, 13, 15, and 17 show those values that are typically obtained from the various alloys. Ranges are shown for most alloys because the properties are frequently controlled by heat treatment and casting process. If desired, a specific requirement can be established between the purchaser and foundry.
3. The tensile properties are measured on cast bars, either end gated or center gated depending on the alloy. These bars are cast with production heats and heat treated with castings. The finished dimensions of these bars will be approximately 1/4" (.6 cm) diameter and 1" (2.5 cm) gage length.
4. Hardness measurements on castings can be used by the foundry to control heat treatment, check for decarburization and to assure consistency of casting practices. The frequency of testing should be as agreed between the purchaser and the foundry.
5. Other special property measurements such as Charpy impact, elevated temperature tensile strength, stress-rupture, corrosion testing, etc., can be performed for critical applications if required. All such requirements must be negotiated by the purchaser and foundry.

Certifications

1. No certifications of chemical, mechanical, or other properties will be provided, unless specifically requested.

Visual Standards

1. Positive Metal - Due to the nature of the investment casting process, random positives are encountered and these vary by alloy and part configuration. Unless otherwise agreed these will be limited to a size of .015-.030" (.038 - .076cm) high by .125" by .125" (.318cm by .318cm), and no more than one per 1" sq. (6.45 square cm). Smaller defects of all sizes may be present at random but not in places where they interfere with the function of the part - such as holes, knurls, or lettering. Positive metal requirements other than this will be specified by the customer at time of quotation.
2. Surface Pits - Random negatives may also occur. These will be limited to .030-.060" (.076 - .152cm), by .030" (.076cm), deep in size, no more than one of those per 1" sq. (6.45 square cm) . Unless otherwise specified by the customer, smaller negatives may be present, providing they do not interfere with the function of the part.
3. Nonfill - Edges may be rounded up to .015" (.038cm), radius even though tooled sharp. Sharper or more rounded edges can be provided as agreed between purchaser and foundry.
4. Cleanliness - Castings will be sand, grit or shot blasted to remove casting and heat treatment scale or other foreign material. Stainless steel castings will not be pickled or passivated unless specified. Castings will be provided with sand blasted, grit blasted or vibro-polished surfaces and be of uniform appearance typical of those processes. Unless the purchaser specifies a rust preventive treatment, some light rusting may be present.
5. Linear Indications - Parts will contain no linear cold shuts, visual cracks, or visual shrinkage, except as agreed with the customer.

Soundness

1. The foundry will establish a casting method that is capable of producing castings that show less than 10% cavity shrinkage porosity in any cross section area. This will be done by visual examination of sectioned castings or radiographic inspection per ASTM E94 during establishment of the casting practice by the foundry. Shrinkage will not be present in an area where subsequent machining will reveal it as a visual surface defect.
2. Magnetic particle inspection per ASTM E-138 or liquid penetrant inspection per ASTM method E-165 will be employed in the sampling stage to make certain the production casting practice is capable of producing crack free castings.

Internal Defects

Some limited quantity of internal defects, including gas holes, slag entrapment, non-metallic inclusions, or porosity may be encountered by the purchaser. If these render the casting unusable, methods of inspection and acceptance limits to exclude such defects will be established by the purchaser and the foundry prior to further production of the casting.

VI. ALLOY TRADEMARKS AND PATENTS

Many alloys in the industry are proprietary developments and registered patents (U.S., U K., etc.). Others have been marketed under registered trademarks. Several owners of such patents or trademarks extract royalties for the use of the name or melting the alloys. This book is intended to be a useful reference and not an authority on the ownership of a particular alloy name or the rights associated with a particular chemistry. A typical example is 17-4 stainless steel, 17-4PH refers to Armco's 17-4PH.

Some examples of common trade designations in this publication are as follows:

Hastelloy Alloys	Illium Alloys
Inconel Alloys	Mar-M Alloys
Nimonic Alloys	

As previously noted, many other trade names are used in this book.

Each Investment Caster, designer or alloy specifier retains responsibility for the proper selection of an alloy in regards to its patented or trademarked rights, or royalties. Special care should be taken accordingly and the company with such claims to an alloy is the best source of information regarding their rights. Other sources of information are the master alloy affiliates, supplier members of the Investment Casting Institute.

Aluminum base casting alloys

The recommended aluminum base alloys for general use are shown in Table 1. These alloys are readily cast, have good fluidity and give accurate reproduction of detail. The most commonly used alloy is A356 because it has excellent castability and good mechanical properties. This alloy is corrosion resistant, readily machinable and can be welded. It can also be anodized and plated.

Where the castings are to be dip brazed alloy D712 is recommended. This alloy is self-hardening and its weldability and machining characteristics are similar to A356 alloy.

Other alloys listed are for special applications. Before specifying any of these alloys, it is suggested that the foundry be given an opportunity to evaluate casting characteristics such as hot shortness, fluidity and surface quality on the new application.

Cobalt base casting alloys

Recommended cobalt-base alloys for general use are shown in Table 3. These alloys are readily cast, have good fluidity and show a minimum of surface imperfections in the cast form. Alloy 21 is the softest, with Alloy J as the hardest within the recommended group.

Alloys listed for special purposes have a variety of uses which range from jet engine blade applications to special cutting tools. Cobalt alloys J, 6 and 12 in the form of cast rod or powders are frequently used in hardfacing applications. Alloy composition and thickness of deposit establish final properties.

The cobalt-chromium-tungsten and/ or molybdenum alloys were originally used for cutting tool, wear and abrasion resistant applications. Further development of these alloys led to the wide use of a Co-Cr-Mo alloy (Vitallium) in dental and surgical implant operations. During 1941, the need for a heat resistant material in highly stressed gas turbine blading led to additional composition modifications and large quantities of blades were produced. This modification and application was an important contribution to the development of many new cobalt-base alloys and founding of the present day investment casting industry.

Early applications of this alloy were at no higher temperatures than those where heat resistant alloys of the Fe-Cr Ni and Fe-Ni-Cr types were used; but the alloy served under conditions of high stress and it excelled in creep and rupture properties. The high temperature cobalt-base alloys contain a moderate amount of carbon and derive their strength by carbide precipitation and from solid solution hardening by chromium, tungsten and molybdenum.

Wear resistant alloys of Co-Cr-W-Mo contain 25 to 32 % Cr and 6 to 20% W or Mo, depending on the application. Nominal carbon contents will vary from 0.60 % for the softer grades to 2.5% for the hard grades. Manganese and silicon are present as deoxidizers; other elements such as vanadium, boron, tantalum, columbium and nickel are added to impart special properties. Unlike steels, the harder grades are generally weaker than the softer grades in tensile and impact strength values.

Outstanding resistance to wear makes these alloys suitable for metal cutting tools and certain machinery parts. Their property of "red hardness"—that is, their ability to retain hardness and strength at high temperatures, gives a superiority over high speed steel in performance and life. Red hardness also makes these alloys more capable of resisting wear where high surface temperatures are present. These alloys have comparatively low coefficients of friction, which means that they develop lower temperatures in sliding contact; therefore, they remain hard. Since they are generally weaker and less ductile than high speed steel, they should not be subjected to extreme conditions of stress that might cause breakage.

The alloys are relatively easily cast with melting usually done in small heats in induction furnaces. No appreciable benefits have been shown by using vacuum or controlled atmosphere melting techniques.

Finishing of castings can be done by machining or grinding practices for the softer grades, whereas the hard grades must be ground. Machining should be accomplished with carbide type tooling. Grinding conditions must be controlled to avoid surface cracking or heat checking with relatively slow wheel speeds desirable.

Copper Base Casting Alloys

Recommended copper base alloys for general use are shown in Table 5.

All copper base casting alloys may be investment cast. Silicon bronze is particularly favored due to its excellent castability, reasonable mechanical properties and good ductility.

Beryllium copper alloy castings were once widely produced but most Investment Casters have abandoned the alloys because of the complex environmental regulations now imposed.

When selecting a copper base alloy, always consult the Investment Caster regarding preferred alloys.

Magnesium base casting alloys

The recommended magnesium base alloys for general use are shown on **Table 7**. Work horse of the industry is AZ91C because it has good castability, high strength-to-weight ratio and excellent machinability.

When brazing is a requirement, alloy AM 100A is recommended. Where higher temperature applications are required, QE 22A alloy can be used.

Other alloys listed are for special purposes. They may require special techniques and may not lend themselves to some applications. It is recommended that purchasers consult the investment casting foundry when applications will require alloys other than those recommended for general use.

Nickel base casting alloys

Recommended nickel base alloy compositions for general use are shown in Table 8. Castability ratings for this group indicate that Ni-Cr-Mo alloys should be selected over the Ni-Mo Fe or the Ni-Cu alloys.

Alloy compositions listed for special purposes have a variety of uses which range from nozzle diaphragms in jet engine applications to parts installed in chemical processing industries.

To insure adequate soundness when casting Monel type alloys, magnesium deoxidation techniques are recommended.

Nickel base alloys, because of their resistance to chemical corrosion and their ability to maintain good mechanical properties at elevated temperatures, are used in a wide range of applications. Because the principal alloying elements vary from one alloy to another, they cannot be placed into a specific classification, but are split into types. These can be identified as Ni-Cr-Mo, Ni-Mo-Fe, and Ni-Cu alloy types.

The Ni-Cr-Mo types are generally identified with the designation "Hastelloy" for which Hastelloy B, C, N and X are common trade names. These alloys contain varying amounts of chromium and molybdenum with nickel as the balance. Hastelloy B, C and N are used extensively where supreme chemical corrosion resistance is required.

Hastelloy B castings are suited for equipment handling hydrochloric acid, hydrogen chloride gas, sulphuric, acetic and phosphoric acids. Hastelloy C can be used in oxidizing and reducing atmospheres up to 2000° F, and is of special interest where parts are highly stressed or subjected to repeated thermal shock within the range of 1600-1800° F. It is one of the few materials which resists the corrosive effects of wet chlorine gas, hypochlorite and chlorine dioxide solutions.

Hastelloy N was developed for use in molten fluoride salts with resistance to aging and embrittlement. Hastelloy X has good high temperature strength and oxidation resistance.

Hastelloy D, which differs only slightly from the Hastelloy alloys mentioned above, finds wide use in the pulp and paper industry for scraper blades, valve bodies and impellers. It can be welded, using techniques similar to cast iron.

All the above alloys have good casting characteristics and may be produced to rigid radiographic and fluorescent penetrant standards. When heat treatment is required, the cycles are easily and readily performed; however, care must be exercised to avoid overheating which would impair tensile strength and corrosion resistance. Welding of these alloys may be readily performed using conventional methods.

The Ni-Cr-Fe types are alloys containing nickel with varying percentages of chromium, iron and minor additions of copper, columbium or silicon. These alloys usually are called "Inconels."

Inconel 610 is a nickel base alloy used for industrial applications that require high strength, pressure tightness and high resistance to destructive chemical corrosion, mechanical wear and oxidation at elevated temperatures. This alloy is not suitable in sulfidizing atmospheres above 1500° F.

Inconel S is a nickel base alloy which is age hardenable with high strength, hardness and resistance to galling. Typical applications are pump components and valves.

The Ni-Cr-Fe alloys have only fair casting characteristics and some difficulty may be encountered in making radiographically sound castings.

The Ni-Cu types are generally identified with the designation "Monel" for which Monel, H Monel and S Monel are common trade names. These alloys are essentially nickel base with varying percentages of copper and silicon. Cast Monel is used for many industrial applications that demand high strength, pressure tightness and high resistance to destructive chemical action or mechanical wear. Although this composition has strength values comparable to the austenitic stainless steels, its corrosion resistance to a wide range of media exceeds that of the stainless steels.

Cast H Monel has a higher silicon content than cast Monel and is used for industrial applications that demand non-galling and anti-seizing characteristics, moderately high hardness and resistance to corrosive attack. In the as-cast condition, strength values will exceed the strength of cast Monel by 10-15%.

Cast S Monel has the highest silicon content of the three alloys in this classification. This alloy is age hardenable. Typical uses include industrial applications that demand high strength, pressure tightness, and high resistance to destructive chemical action and galling; it is particularly suitable for valve seats and sliding or other moving parts.

Care must be exercised in casting these alloys so that integral soundness may be obtained to pass necessary radiographic and/or fluorescent penetrant standards. To achieve these requirements it is usually necessary to control pouring temperature, height of pour and to deoxidize with magnesium. These types of Monel have only fair casting characteristics necessitating good foundry practices.

Beryllium nickel

Castings of nickel-base alloy with 2 to 3% Be exhibit many of the properties of hot worked die steel. The Be addition improves castability and imparts a steel-like hardness. Cast Be-Ni alloys can be hardened to Rc 52-55; they retain this hardness along with oxidation resistance up to 950° F.

Excellent wear characteristics of Be-Ni are due to the hard nickel berylides which act like carbides in Cr-Mo-V and Cr-W-V steels. The cast alloy resists heat checking and crazing, is stainless and develops a surface oxide film which adds wear resistance. Cryogenic properties are excellent.

Beryllium promotes good feeding and sound castings with simple risering. The alloy is ideal for intricate investment castings. It can be air melted in furnaces with dense magnesia linings. Deoxidation with Mg or Si should precede the Be-master alloy addition.

Melting range is 2100° - 2300° F, pouring range is 2500° - 2600° F depending on size and detail of the casting and on mold temperature. Annealing is accomplished at 1950° F for two to three hours, followed by a water or oil quench.

In this condition, beryllium nickel machines like Udimet 500 or Rene 41.

Be-Ni alloys are used for plastic injection molds, casting dies, spark-resistant tools, diaphragms, surgical instruments and corrosion resistant components.

Carbon and low alloy steels

Recommended carbon and low alloy steel compositions for general use are shown in Table 10. Note that silicon content has been raised to 1.00% maximum for deoxidation and castability. Other alloys for special purposes are also listed with the same modification.

Care should be exercised when processing parts made from these alloys because they are subject to surface decarburization, therefore necessitating a carbon restoration treatment. This treatment will produce uniform carbon content throughout the parts and, when subsequently processed through heat treating cycles, uniform mechanical and hardness values will result.

Carbon and low alloy steel castings, because they can be processed to obtain a wide range of properties, find many uses in commercial applications. The current availability of these alloys in conjunction with modern heat treating techniques can give the designer much freedom. Steel castings are usually purchased to meet specified mechanical properties, therefore, some leeway is permitted in chemical composition.

This leeway in chemical composition permits the caster to use compositions which reduce problems in the casting process. Since soundness is essential to integrity of a casting, it is advisable not to specify manganese and silicon to a range, but to permit the caster freedom of choice for these elements to achieve soundness in the shape being cast. The silicon content is preferably higher in cast steels, however, a content in excess of 0.80% it is considered an alloy addition since it imparts resistance to tempering.

Carbon and low alloy steel castings within this section are usually divided into four general groups. These are (1) low carbon steel castings with less than 0.20% C, (2) medium carbon steel castings with 0.20-0.50% C, (3) high carbon steel castings with more than 0.50% C and (4) low alloy steel castings containing a maximum of 8% alloy content.

Carbon steels

Low carbon steels are frequently specified by SAE designation for low cost parts. Government or ASTM casting specifications, however, allow broader ranges in chemical composition, which permits more control over element losses. SAE designations are wrought compositions.

Initial unit cost of carbon steels has far less effect on cost of the finished part than the castability or fluidity of the molten steel. By substituting a low or high alloy steel (at slightly greater cost), superior

79

castability, a better cast surface finish and, consequently, less mate-
rial loss in finishing can be achieved.
Many casters have established that it usually costs less to cast 8620
low alloy steel than a 1020 composition. Alloy steel 8620, through
heat treatment, yields a higher recovery, provides better physical and
wear resisting properties and a smoother surface finish.

Low alloy steels

Low alloy cast steels comprise two groups according to their end use: those for structural parts of increased strength, hardenability and toughness and those resistant to wear, abrasion or to heat and corrosive attack. Both groups can be used as described in the appropriate AMS specifications.

Cast low alloy steels have been developed to operate under conditions requiring greater pressures, wear resistance, higher impact resistance and higher strength with increased toughness and hardenability. Tensile strength values, depending on heat treatment and composition, will range from 70,000 to 200,000 psi. Based on these characteristics, a trend has developed to design with low alloy steel compositions for lighter weight with no loss in strength.

An example of alloy substitution for particular properties would show that, for highly stressed parts, the 4100 series of low alloy steels is preferred to carbon steels of comparable carbon content. Composition 4130 is one of the best shock resistant investment casting ferrous alloys. For higher strength, but with some sacrifice of impact resistance, 4140 or 4150 may be used.

These chromium-molybdenum steels (41xx) are generally recommended as substitutes for all other 0.40 or 0.50% carbon steels such as 1040, 3140, 4640, 6140 or 8640.

Although relatively high impact properties, hardness and strength are obtained by using 6150, wear and abrasion resistance is usually lower than with 52100 or 1095 steels. Alloy 52100 should be specified in place of 1095, because it has better castability, wear and abrasion resistance. Both steels have low impact values when hardened above Rockwell C 45.

Cast stainless steels
Corrosion, heat resistant and precipitation hardening stainless steel alloys

Corrosion and heat resistant castings are distinguished by their ability to serve where carbon and low alloy steels would be destroyed by the action of the environment rather than by mechanical conditions of loading.

Recommended alloys for most general applications covering a wide range of mechanical properties are shown in tables 12-15. These alloys are readily cast, usually have good fluidity and resistance to hot tearing, and are easily heat treated to desired mechanical properties.

A knowledge of the basic characteristics of the 300 and 400 series stainless steels can be a guide to correct application of these alloys. Stainless steels are essentially low carbon alloy steels containing a minimum of 11.5% Cr. The chromium addition provides resistance to corrosion and to scaling at elevated temperatures. Carbon contents are held to 0.20% or less, except in some types requiring high hardenability for applications such as bearings and fine cutlery. In such cases, the increased carbon demands increased chromium to maintain stainless properties. Other elements such as copper, nickel, molybdenum, columbium/tantalum, aluminum, sulphur and selenium are added to produce special properties.

Stainless steels are selected for their high strength, abrasion and erosion resistance, magnetic properties, their attractive appearance and the ease of cleaning their smooth dense surfaces.

There are three main classes of stainless steels; the straight chromium hardenable 400 series martensitic types; the straight chromium non-hardenable 400 series ferritic types; and the austenitic types of chromium-nickel, 300 series.

Hardenable Types:

Martensitic steels of the 400 series are capable of being heat treated to a wide range of useful hardness and strength levels. They contain chromium as the major alloying element. Chromium and carbon contents are balanced so that the soft austenitic phase, which develops at high temperatures, transforms to a hard martensitic phase during cooling to room temperature.

Carbon content is usually under 0.15% when chromium is 11.50-14.00%, as in types 403, 410 and 416. Higher carbon content up to 1.20 % is used with chromium levels of 16.00-18.00% as in the 440 types. The martensitic stainless steels provide hardness values up to about Rockwell C 62 and tensile strengths to about 285,000 psi. They are magnetic in all conditions and resist wear and abrasion when used for steam and gas turbine parts, for cutlery and bearings. These types are not as corrosion resistant as the ferritic and austenitic types.

A small group of alloys known as the "Precipitation Hardening Alloys" would come under the hardenable types of martensitic stainless steels, however, final properties are achieved through solution treating followed by an aging cycle. Chemical composition will vary somewhat; however, they contain approximately 16% Cr, varying percentages of nickel, copper or molybdenum, depending on alloy, with a low carbon content. Strength levels can range from 130,000 to over 180,000 psi tensile, hardness values in the range of Rockwell C 33-42, with good ductility.

Non-Hardenable Types:

The ferritic steels of the 400 series are normally used for their corrosion resistance and resistance to scaling at elevated temperatures rather than for high strength purposes. The ferritic condition is maintained by a chromium-to-carbon balance which suppresses the formation of austenite. These grades contain low carbon and relatively high chromium, ranging from 14.00 to 18.00% Cr in type

430 to a high of 23.00 to 27.00% Cr in type 446. Increasing chromium results in increased corrosion resistance and resistance to scaling, but with some sacrifice in other properties, such as impact strength.

The cast ferritic grades have rather low tensile values of approximately 75,000 psi, ductility of approximately 10-15% and relatively low hardness values, usually Rockwell B 100 maximum.

Austenitic Types:

The addition of nickel as the second major alloying element produces austenitic stainless steels of the 300 series. These types retain an austenitic structure during cooling from elevated temperatures. The basic type, 302, contains 18.00% Cr and 8.00% Ni. More or less chromium and nickel, as well as other alloying elements, are added for special purpose types.

The austenitic types have high ductility, low yield strength and tensile strengths comparable to the ferritic stainless grades. They are readily welded and mechanical properties may be improved by cold working. The austenitic types are non-magnetic as annealed and, depending on composition, may become slightly magnetic when cold worked. They have excellent properties at cryogenic temperatures and surpass the strength of any 400 series type at temperatures above 1000° F.

Austenitic stainless provides the best corrosion resistance of all the stainless steels, particularly when they have been annealed to dissolve chromium carbides and then rapidly quenched to retain carbon in solution. The problem of selecting an alloy for corrosive conditions involves the following factors:

1. Method of casting.
2. Availability of metals which would solve the problem.
3. The type of corrosion and the allowable rate of attack—which may be divided into three classes:
   a) Conditions in which color, taste or odor of the final product are of importance as in the food, beverage, dairy and chemical industries.
   b) Ornamental parts where freedom from discoloration is vital.
   c) Components in industry which must remain physically intact over economical periods of service operation and in which the rate of corrosion and the life are important.

Heat resistant cast stainless alloys

The high inherent hot strength of Ni-Cr-Fe alloys precludes forging or rolling; such alloys are readily cast. Alloy selection for high temperature service is influenced by corrosive conditions and the range of expected service temperatures.

Surface stability in hot corrosive gases will vary with alloy composition, since none are immune to attack but their corrosion is sufficiently slow for allpractical uses. Chart 2 indicates the approximate resistance of cast high alloys to various atmospheres at 1800° F for a 100-hour period.

Structural stability or resistance to dimensional change under stress and temperature is important in selection of an alloy for high temperature service. This property is related to time, stress and the life of the part which must be considered at the outset. A designer should not assume that strength data based on laboratory data can be applied arbitrarily to a particular design application. Allowable design stresses should be selected with due consideration for conditions of loading and should contain an "experience factor" which can be obtained by consulting the design department of a high alloy foundry.

CHART 2						
RESISTANCE OF CAST HEAT-RESISTANT ALLOYS TO CORROSION IN VARIOUS ATMOSPHERES AT 1800°F FOR A 100-HOUR PERIOD						
Alloy	Corrosion in Air	Corrosion in Oxidizing Flue gas 5 gs	Corrosion in Reducing Flue gas 5 gs	Corrosion in Reducing Flue gas 300 gs	Corrosion in Reducing Flue Gas (const.temp.) 100 gs	Corrosion in Reducing Flue Gas (cooled to 300°F each 12 hrs.) 100 gs
HC	G	G	G	S	G	G
HD	G	G	G	S	G	G
HE	G	G	G	-	G	-
HF	S	G	S	U	S	S
HH	G	G	G	S	G	G
HK	G	G	G	U	G	G
HT	G	G	G	U	S	U

G—Good, S—Satisfactory, U—Unsatisfactory, gs—Grains sulfur per 100 cu ft of gas.

Corrosion

Corrosion, in general, is the destructive attack of a metal, elemental or alloyed, by chemical or electrochemical reaction with its environment. No two metals react identically in a given environment. Although this definition is general, we can list four major types of corrosion which are of concern to design engineers. These are:

1. Galvanic Corrosion
2. Stress Corrosion
3. Corrosion Fatigue
4. Erosion Corrosion

Galvanic corrosion is corrosion associated with the current of a galvanic cell, consisting of two dissimilar conductors in an electrolyte or two similar conductors in dissimilar electrolytes. Where the two dissimilar metals are in contact, the resulting reaction is referred to as "couple action."

All metallic materials fit into a galvanic series, Chart 3 - Relative position in this table determines, in part, the cathodic or anodic relationship of materials. In general, the cathodic or noble metals resist attack and the anodic metals suffer.

While corrosion is a complex phenomenon, the fundamental reaction in galvanic corrosion involves a transfer of electrons: some positively charged ions in the corroding solution—usually hydrogen ions—lose electrical charges which are required by the metal or alloy, which goes into solution, or corrodes. The corrosion reaction is divided into an anodic portion and a cathodic portion occurring at discrete points on metallic surfaces. Anodic reaction (oxidation) represents the acquisition of charges by the corroding metal, while cathodic reaction (reduction) represents neutralization by hydrogen ions, which are discharged. The flow of electricity between anodic and cathodic areas may be generated by local cells set up either on a single metallic surface or between dissimilar metals.

Prime factors which determine the extent or progress of corrosion include:

a)   Acidity of the solution
b)   Oxidizing agents
e)   Temperature
d)   Effect of films
e)   Inhibitors
f)   Metal - ion concentration cells
g)   Oxygen concentration cells

Other factors which affect corrosion to varying degrees are (a) surface condition, (b) residual and operating stresses and (c) heat treatment and welding.

Based on experiences in corrosion testing and a knowledge of the galvanic behavior of metals and alloys, Chart 3 indicates the tendency of metals and alloys to form galvanic cells and to predict the probable direction of the galvanic effect. This series is not to be confused with the "Electromotive Series." The Galvanic Series takes into consideration overall and practical aspects in addition to theoretical principles. The further apart metals stand, the greater will be the galvanic tendency. This can be determined by measuring the electric potential difference between them.

Stress corrosion is the destructive attack of a metal or alloy while under the influence of stress. It leads to stress corrosion cracking, which indicates combined action of static tensile stress and corrosion; these lead to cracking. Principal factors involved are stress environment, time and internal structure of the alloy. These factors interact, one accelerating the action of another and their relative importance varies.

If stress corrosion cracking is to occur, there must be tensile stresses at the surface. The stress may be internal or applied. Internal stress may be produced by unequal cooling from high temperature, internal structural rearrangements involving volume changes, welding and straightening and/or coining operations. These concealed stresses are often of greater importance than actual operating stresses.

One curious aspect of stress corrosion cracking is the wide difference in time required for failure, which varies from a matter of minutes to many years. A long time may pass before corrosion proceeds to such an extent that it begins to be accelerated by tensile stresses present. More severe corrosive conditions and higher initial stress promote earlier cracking.

Corrosion fatigue can be defined as the effect of applying repeated or fluctuating stresses in a corrosive environment. It is characterized by shorter part life than would be encountered as a result of either repeated or fluctuating stresses alone or corrosive environment alone. In static environments, corrosion products and films tend to block or retard the attack—even cause it to cease completely. Cyclic stresses, however, tend to rupture the film or render it more permeable. As a result, cyclic stresses accelerate corrosion. Notches also accelerate the effect of cyclic stress. Surface roughness or pitted surfaces—as-cast or formed by corrosion—will form notches, concentrate the stress and accelerate corrosion. These mutually accelerating processes continue until failure occurs by corrosion fatigue.

In acid solutions, where the possibilities of film formation are less likely, the mechanism may be somewhat different. Cyclic stresses may produce distorted metal which is more susceptible to corrosion or it may change the area ratios and polarization characteristics of local anodes and cathodes. In any case, the effect of cyclic stresses is to increase corrosion and to cause the corrosion to form deep notches.

Of the two accelerating factors, the effect of cyclic stresses on corrosion is the more important. All the factors which influence normal corrosion have some effect on corrosion fatigue. Hence, composition of the corroding medium, temperature, degree of aeration, and velocity must be carefully controlled. Also, the possibilities of galvanic effects should be considered.

Erosion Corrosion is the destruction of metal or other materials by the abrasive action of moving fluids, usually accelerated by the presence of solid particles or matter in suspension.

At first glance, erosion may seem to have little to do with corrosion; however, many metals owe their corrosion resistance to protective films. These may be oxide films, layers of corrosion products, or some other type of protection. Removal of these protective films by erosive action exposes fresh metal to corrosion attack. As a result, the metal corrodes much faster than in the absence of erosion. Conditions typical of erosion corrosion involve liquids moving at high velocities solids in suspension, marked turbulence and impingement. All exposure conditions that determine corrosion rates under normal conditions have some effect on erosion corrosion because they influence the nature and rate of formation of protective films. Alloying, which increases corrosion resistance and develops more protective films, will be helpful in erosion corrosion.

As a companion to erosion corrosion, cavitation erosion is a related action. Cavitation erosion may occur anywhere a liquid is in contact with a metal surface and subjected to rapidly alternating ranges of pressure. At high relative motion between metal and liquid, pressure on the liquid may be reduced locally to the boiling point. At these extreme low pressure regions, small cavities of vapor form in the liquid. Return of the pressure to normal causes an implosion as the cavity collapses.

If the cavity is in contact with the metal, there is a high speed of liquid impact. The surface then work hardens, roughens and cracks by fatigue. Deep pits may also form which then make the surface spongy. Formation of erosion pits or any roughening of the surface by corrosion increases turbulence at the moving metal surface and thereby increases the danger of cavitation. Conditions may be such that a corrosion resistant material might remain unaffected, whereas a material susceptible to pitting would soon develop accelerated damage due to cavitation-erosion effects. The corrosion may also lower fatigue resistance.

CHART 3
GALVANIC SERIES OF METALS AND ALLOYS

Corroded End (anodic, least noble)

1.	Magnesium	13.	Tin	25.	Chromium-Iron (passive)
2.	Magnesium Alloys	14.	Nickel (active)	26.	Titanium
3.	Zinc	15.	Inconel (active)	27.	18-8 Cr-Ni Iron (passive)
4.	Aluminum 25	16.	Hastelloy C (active)	28.	Hastelloy C (passive)
5.	Cadmium	17.	Brasses	29.	Silver
6.	Aluminum	18.	Copper	30.	Graphite
7.	Steel or Iron	19.	Bronzes	31.	Gold
8.	Cast Iron	20.	Copper-Nickel Alloys	32.	Platinum
9.	Chromium Iron (active)	21.	Monel		**Protected End** (cathodic,
10.	18-8 Cr-Ni Iron (active)	22.	Silver Solder		most noble)
11.	Lead-Tin Solders	23.	Nickel (passive)		
12.	Lead	23.	Inconel (passive)		

Chromium-iron and chromium-nickel-iron alloys frequently change positions as indicated. Inconel and sometimes nickel, behave in a similar manner. This depends upon the corrosive medium; its oxidizing power and acidity or the presence of activating ions, such as halides.

Tool and die steel casting alloys

Recommended tool and die steel compositions for general use are shown in Table 16. Note that silicon content is considerably higher than in wrought products of comparable composition. Other alloys listed for special purposes have a similar modification.

Most of the alloys listed have similar casting characteristics. Alloy cost and end use were given major consideration. Since heat treating systems vary from plant to plant, controlled atmosphere, salt or vacuum systems should be used during hardening cycles. In the event excessive surface decarburization is present, carbon restoration processes may be used.

There are few casting specifications for this group of steels; their use as investment castings is specialized. Cast tool steels use the wrought, SAE, nomenclature EXCEPT that a letter "C" precedes the name. This is necessary because the chemical specifications for cast tool steel often differ from the wrought counterpart. They will be supplied to meet chemical analysis ranges as shown in Table 16. Standard Hardness values only are listed for mechanical properties, Table 17.

Tool Steels are characterized by high hardness and resistance to abrasion coupled, in many instances, with resistance to softening at elevated temperature.

Die Steels, although having similar characteristics, are usually used at lower hardness ranges where they have higher impact resistance.

To simplify classification and selection of tool and die steels, a system has been developed by the American Iron and Steel Institute and the Society of Automotive Engineers. Some 95 chemical compositions are classified into 14 groups. Although many are castable and used for special purposes, recommended alloys for general use are:

1. Air-hardening medium alloy cold work steel—A-2 designation
2. High carbon high chromium cold work steel—D-2 designation
3. Chromium hot work steel—H-11 designation
4. Molybdenum high speed steel—M-2 designation
5. Oil hardening cold work steel—O-1 designation
6. Shock resisting steel—S-1 designation
7. Alloys 440C and 52100, although not classified as tool steel, may be used in cutting applications.

Characteristics of the designations above are as follows:

Air-hardening medium alloy cold work steels contain manganese, chromium, molybdenum and vanadium as the alloying elements. This group is deep hardening, so it has air-hardening characteristics with low distortion. High carbon content insures high wear resistance with considerable resistance to heat softening. Casting characteristics are good, however, care must be exercised to avoid excessive decarburization. Heat treating cycles are relatively simple with hardness values of Rc 58-62 as a desired working range.

High carbon high chromium cold work steels contain carbon and chromium as major alloying elements with molybdenum as a minor alloying element. Some compositions have tungsten, cobalt and vanadium as optional addition elements. These steels are highly wear resistant with deep hardening promoted by the high carbon and chromium contents. Hardenability is accentuated by minor additions of tungsten and molybdenum. This careful balance of alloying elements and air-hardening properties results in extremely low dimensional change during heat treatment.

Medium resistance to heat softening, however, limits the use of this group to applications below 900° F. Susceptibility to edge brittleness in combination with relatively low heat resistance makes this alloy unsuitable for cutting tools. Typical uses are long-run blanking and forming dies, brick molds, gages and abrasion-resistant liners.

Castability characteristics are excellent; however, care must be exercised to avoid excessive surface decarburization. Heat treating cycles are relatively simple with hardness values of Rc 58-64 as the desired working range.

Chromium hot work steels contain chromium and tungsten with additions of molybdenum and vanadium. They have good resistance to heat softening because of their medium chromium content supplemented by addition of the carbide-forming molybdenum, tungsten or vanadium. Low carbon and relatively low total alloy content promote toughness at the normal working hardness of Rc 40-55. Higher tungsten and molybdenum contents increase red hardness and hot strength, but will slightly reduce toughness.

The steels are extremely deep hardening and may be air hardened in heavy sections. This property, with balanced alloy content, is responsible for low distortion in hardening. Typical uses include forging dies, die casting dies, mandrels, etc. Castability characteristics may be rated as good.

Molybdenum high speed steels contain molybdenum, tungsten, chromium, vanadium, cobalt and carbon as major alloying elements. These steels are similar in properties to the tungsten high speed tool steels, but generally have slightly greater toughness at the same hardness. The main advantage of this group over the tungsten group lies in their lower cost, while maintaining equivalent performance. Increasing carbon and vanadium content increases wear resistance; increasing cobalt raises red hardness.

The molybdenum high speed steels are sensitive to hardening conditions, particularly to temperature and atmosphere, since they will decarburize and overheat easily under adverse treating conditions. Typical uses are for cutting tools of all types.

Castability characteristics are good, however, care must be exercised to avoid excessive surface decarburization. With proper precautions as to time and temperature, hardening and tempering is relatively simple with normal working hardness of Rc 64-66. Normal hardening cycles are conducted in salt baths followed by hot salt quenching and air cooling.

Oil hardening cold work steels contain tungsten, manganese, chromium and vanadium as an optional element. Alloy additions increase the hardenability, permitting oil quenching, with much less distortion and less cracking hazard than with the water hardening group. This group is relatively inexpensive and its high carbon content produces adequate wear resistance for short-run applications at or near room temperature. Typical uses for these steels are short-run cold forming dies, blanking dies and gages where distortion is unimportant and cutting tools where no high temperatures are generated.

Since wear resistance of cold work die steels is an important factor, carbon contents of this group are high. This results in considerable undissolved carbide at the hardening temperature so despite their relatively high alloy contents, some of the grades in this group do not harden as deeply as the lower alloy shock resisting steels. Raising the hardening temperature has several effects on hardenability. It increases grain size and increases solution of alloying elements with subsequent increase in hardenability when oil quenched.

Castability characteristics are good, however, care must be exercised to avoid excessive surface decarburization. Hardening and tempering cycles are relatively simple with a normal wording hardness range of Rc 58-62.

Shock resisting steels contain silicon, chromium, tungsten, nickel and sometimes molybdenum as alloying elements. Silicon and nickel strengthen the ferrite and increase hardenability; chromium improves hardenability, provides some heat resistance and contributes slightly to abrasion resistance. With carbon content maintained at about 0.50%, these steels have high strength with moderate wear resistance. Tools made from these steels have measurable ductility even at Rc 60. Principal uses are for chisels, rivet sets, hammers and other tools where repetitive high-impact loading is developed.

Hardenability will vary depending on composition within this group; however, the recommended composition, S-l, is deep hardening. Grade S-1 is normally oil hardened followed by tempering to a hardness range of Rc 53-58.

Castability characteristics are good, however, care must be exercised to avoid excessive surface decarburization. Hardening and tempering cycles are relatively simple.

440C and 52100 are high carbon-high chromium and high carbon-low chromium alloy compositions. Both alloys are usually not classified as tool and die steels, but they have been included for their high hardenability and wear resistance. Alloy 440C may be hardened by quenching in either air or oil, whereas52100 is usually water or oil quenched. Of the two alloys, 440C has excellent corrosion resistance.

Castability of these alloys may be rated as good when care is taken to avoid excessive surface decarburization. Heat treating procedures are relatively simple with normal hardness ranges being Rc 54-68 for 440C and Rc 59-63 for 52100.

TABLE 1 - ALUMINUM ALLOYS
Typical Chemical Range Percentages

Trade Name	Cu	Si	Mg	Ti	Fe	Mn	Zn	Cr	Other	Trace Ea.	Trace Tot.
Pure Aluminum	0.03	0.2	0.03	-	0.2	0.03	0.03	-	-	0.03	0.05
D-712(40E)	0.25	0.3	.50-.65	.15-.25	0.5	0.1	5.0-6.5	.40-.6	-	0.05	0.2
RR350	4.50-5.50	0.2	-	.15-.25	0.3	.20-.30	-	-	Ni - 1.3- 1.8 Co - .10- .40 Zr. - .10- .30	0.05	0.3
354	1.6-2.0	8.6-9.4	.40-.6	0.2	0.2	0.1	0.1	-	-	0.05	0.15
355	1.0-1.5	4.5-5.5	.40-.6	0.25	0.6	0.5	0.35	0.25	-	0.05	0.15
C-355	1.0-1.5	4.5-5.5	.40-.6	0.2	0.2	0.1	0.1	-	-	0.05	0.15
356	0.25	6.5-7.5	.20-.40	0.25	0.6	0.35	0.35	-	-	0.05	0.15
A-356	0.2	6.5-7.5	.20-.40	0.2	0.2	0.1	0.1	-	-	0.05	0.15
357	0.5	6.5-7.5	.45-.6	.10-.20	0.15	0.3	0.3	-	-	0.03	0.09
A-201 (KO-1)	4.0-5.0	0.05	.18-.35	.15-.35	0.05	.20-.30	-	-	Ag. .40-1.0	0.03	0.1
Precedent- 71A	0.1	0.15	.8-1.0	.10-.20	0.15	0.1	6.5-7.5	.06-.20	-	0.05	0.15

*   Ti + Zr = .50
** .10- .40Sb, Sb + Co = 9.6
1   Where two numbers are not shown in a block, the value is a maximum. This applies to all tables of chemistry unless specifically noted

TABLE 2
PROPERTIES OF SEPARATELY CAST TEST BARS
OF ALUMINUM BASE ALLOYS

Alloy	Tensile Strength English psi	Tensile Strength Metric MPa	0.2% Yield Strength English psi	0.2% Yield Strength Metric MPa	% Elongation Range (in 2.5cm)
356	32-40,000	221-276	22-30,000	152-207	3-7
A-356	38-48,000	262-331	28-36,000	193-248	3-10
A-357	33-50,000	228-345	27-40,000	186-276	3-9
355 C-355	35-50,000	241-345	28-39,000	193-269	1-8
D-712 (40E)	34-40,000	234-276	25-32,000	172-221	4-8
A-354	47-55,000	324-379	36-45,000	248-310	2-5
RR-350	32-45,000	221-310	24-38,000	165-262	1.5-5
Precedent 71	35-55,000	241-379	25-45,000	172-310	2-5
A-201 (KO-1)	56-60,000	386-414	48-55,000	331-379	3-5

NOTE: The above mechanical property values are for information only. They do not necessarily apply to castings. Any requirements for mechanical properties are beyond this standard and must be negotiated with the foundry.

TABLE 3 - HIGH COBALT ALLOYS
Chemical Ranges to be Provided (%)

Trade Name	C	Mn	P	S	Si	Ni	Cr	Mo	Fe	Co	W	Other
Cobalt J	2.2 2.7	1.00	0.03	0.03	1.00	2.50	31.0 34.0	-	3.00	Bal.	16.0 19.0	.25B, 2.0 Total
Cobalt 3	2.0 2.7	1.00	0.03	0.03	1.00	3	29.0 33.0	-	3.00	Bal.	11.0 14.0	2.0 Total
Cobalt 6	.9 1.4	1.00	-	-	1.50	3.00	27.0 31.0	1.50	3.00	Bal.	3.5 5.5	
Cobalt 12	1.10 1.70	1.00	0.03	0.03	1.00	3.00	28.0 32.0	-	3.00	Bal.	7.00 9.50	
Cobalt 19	1.5 2.0	1.00	0.03	0.03	1.00	-	29.0 33.0	-	3.00	Bal.	9.0 12.0	
Cobalt 21	.20 .30	1.00	0.04	0.04	1.00	1.75 3.75	25.0 29.0	5.0 6.0	3.00	Bal.		.007B
Cobalt 25	.05 .15	1.0 2.0	0.04	0.04	1.00	9.0 11.0	19.0 21.0	-	3.00	Bal.	14.0 16.0	
Cobalt 31	.45 .55	1.00	0.04	0.04	1.00	9.5 11.5	24.5 26.5	-	2.00	Bal.	7.0 8.0	
Cobalt 36	.35 .45	1.0 1.5	0.03	0.03	0.35	9.0 11.0	17.5 19.5	-	2.00	Bal.	14.0 15.0	.01-.05B
Cobalt 93	2.75 3.25	1.50	0.03	0.03	1.50		15.0 19.0	14.0 18.0	Bal	4.0 7.0		1.5-2.5V
N-155	0.20	1.00 2.00	0.04	0.03	1.00	19.0 21.0	20.0 22.5	2.50 3.50	Bal.	18.5 21.0	2.00 3.00	.10-0.20N,0.75-1.25 Cb + Ta
Tantung G	1.8 2.2	-	-	-	-	-	26.0 29.0	-	2.00	Bal.	15.0 17.0	.15-.25B; 4.5-5.5 Ta
WI-52	.40 .50	0.50	0.04	0.04	0.50	1.00	20.0 22.0	-	1.0 2.5	Bal.	10.0 12.0	1.5-2.5 Cb + Ta
F75	.20 .35	1.00	-	-	1.00	2.50	27.0 30.0	5.0 7.0	0.75	Bal.	-	-

TABLE 4 - PROPERTIES OF SEPARATELY CAST TEST BARS OF COBALT BASE ALLOYS

Alloy	Condition	Tensile Strength		0.2% Yield Strength		% Elongation Range (in 2.5 cm)	Hardness Rc Range
		English psi	Metric MPa	English psi	Metric MPa		
J	As Cast						55-60
3	As Cast						48-53
6	As Cast						37-45
12	As Cast						44-50
19	As Cast						47-52
21	As Cast	95-130,000	655-896	65-95,000	448-655	8-20	24-32
25	As Cast	90-120,000	621-827	60-75,000	414-517	15-25	20-25
31	As Cast	105-130,000	724-896	75-90,000	517-621	6-10	20-30
36	As Cast	90-105,000	621-724	60-70,000	414-483	15-20	30-36
93	As Cast						61-65
N-155	Solution Anneal	90-100,000	621-690	50-60,000	345-414	15-30	90-100 (Rb)
Tantung G							48-53
WI-52	As Cast	90-105,000	621-724	65-75,000	448-517	5-9	32-38
F75	As Cast	95-110,000	655-758	70-80,000	483-552	8-15	25-34

NOTE: *The above mechanical property values are for information only. They do not necessarily apply to casting. Any requirements for mechanical properties are beyond this standard and must be negotiated with the foundry.*

TABLE 5 - COPPER BASE ALLOYS
Typical Chemical Range Percentages

Alloy & (CDA No.)	Al	Zn	Sn	Pb	Ni	Fe	Si	Be	Co	Mn	P	Cr	Cu	Total Others
Aluminum Bronze A (952)	8.5 9.5	--	--	--	--	2.5 4.0	--	--	--	--	--	--	86.0[1]	1.0
Aluminum Bronze B (953)	9.0 11.0	--	--	--	--	.75 1.5	--	--	--	--	--	--	86.0[1]	1.0
Aluminum Bronze C (954)	10.0 11.5	--	--	--	2.5	3.0 5.0	--	--	--	.50	--	--	83.0[1].	50
Aluminum Bronze D (955)	10.0 11.5	--	--	--	3.0 5.5	3.0 5.0	--	--	--	3.5	--	--	78.0[1]	.50
BeCu 10C (820)	.10	.01	.01	.01	.20	.10	.15	.45 .8	2.4 2.7[2]	--	--	.01	Bal.	--
BeCu 20C (825)	.15	.10	.10	.02	.20	.25	.20 .35	1.9 2.2	.35 .7[2]	--	--	.10	Bal.	--
BeCu 165C (824)	--	--	--	--	--	--	--	1.7 1.8	.20 .30	--	--	--	97.20[1]	--
BeCu 275C (828)	.15	.10	.10	.02	.20	.25	.20 .35	2.5 2.8	.35 .7[2]	--	--	.10	Bal.	--
Pure Copper	--	--	--	--	--	--	--	--	--	--	--	--	99.95[1]	--
Chrome Copper	--	--	--	.015	--	.05	.10	--	--	--	--	.4 1.0	Bal.	.1
Leaded Yellow Brass (854)	.35	24.0 32.0	5.0 1.5	1.5 3.8	1.0	.7	.05	--	--	--	--	--	65.0 70.0	
Red Brass (836)	--	4.0 6.0	4.0 6.0	4.0 6.0	1.0	.30	--	--	--	--	.05	--	84.0 86.0	
Manganese Bronze A (865)	.50 1.5	36.0 42.0	1.0	.40	1.0	.40 2.0	--	--	--	.10 1.5	--	--	55.0 60.0	
Manganese Bronze C (863)	5.0 7.5	22.0 28.0	.20	.20	1.0	2.0 4.0	--	--	--	2.5 5.0	--	--	60.0 66.0	
Tin Bronze (903)	--	3.0 5.0	7.5 9.0	.30	1.0	.15	--	--	--	--	.05	--	86.0 89.0	
Manganese Copper 2.75	--	--	--	--	--	--	--	--	--	2.5 3.0	--	--	Bal.	.15
Manganese Copper 3.5	--	--	--	--	--	--	--	--	--	3.35 3.75	--	--	Bal.	.15
Manganese Copper 5.0	--	--	--	--	--	--	--	--	--	4.75 5.25	--	--	Bal.	.15
Silicon Brass (875)	.50	12.0 16.0	--	.50	--	--	3.0 5.0	--	--	--	--	--	79.0[1]	--
Silicon Bronze (872)	1.5	5.0	1.0	.50	--	2.5	1.0 5.0	--	--	1.5	--	--	89.0[1]	--

[1] - Minimum, others where no range is shown are maximum.

[2] - Ni + Co

TABLE 6 - PROPERTIES OF SEPARATELY CAST TEST BARS
OF COPPER BASE ALLOYS

Alloy - (CDA No.).		Tensile Strength		Yield Strength		% Elongation	Hardness(Rb)
		English psi	Metric MPa	English psi	Metric MP	Range (in 2.5cm)	Range
Al.Bronze C	A.C.	75-85,000	517-586	30-40,000	207-276	10-20	80-85
(954)	H.T.	90-105,000	621-724	45-55,000	310-379	6-10	91-96
Al.Bronze D	A.C.	90-100,000	621-690	40-50,000	276-347	6-10	91-96
(955)	H.T.	110-120,000	758-827	60-70,000	414-552	5-8	93-98
Tin Bronze		40-50,000	276-347	18-30,000	124-207	20-35	40-50
Red Brass		30-40,000	207-276	14-25,000	97-172	20-30	30-35
MnBr, A		65-75,000	448-517	25-40,000	172-276	16-24	60-65
MnBr C		110-120,000	758-827	60-70,000	414-483	8-16	95-100
BeCu 10C (820)	A.C.	45-50,000	310-347	20-25,000	138-172	15-20	50-55
	Hard.	90-100,000	621-690	50-60,000	347-414	3-8	90-95
BeCu 20C (825)	A.C.	70-80,000	483-552	40-45,000	276-310	18-23	75-80
	Hard.	110-160,000	758-1103	90-130,000	621-896	1-4	25-44(Rc)
BeCu 275C (828)	A.C.	80-90,000	552-621	50-55,000	347-379	15-20	80-85
	Hard	---	---	---	---	---	42-46(Rc)
Copper		20-30,000	138-207	---	---	4-50	35-42
Cr Copper		33-50,000	228-347	20-40,000	138-276	20-30	70-78
BeCu 165C (824)		70-155,000	483-1069	40-140,000	276-965	1-15	60(Rb) - 38 (Rc)
Leaded Yellow Brass (854)		30-50,000	207-347	11-20,000	76-138	15-25	---

NOTE: The above mechanical property values are for information only. They do not necessarily apply to castings. Any requirements for mechanical properties are beyond this standard and must be negotiated with the foundry.

* Yield strength is determined by 0.5% extension under load or 0.2% offset method.

A.C. = as cast   H.T. = heat treated

TABLE 7 - MAGNESIUM ALLOYS

Designation of Alloys Recommended for General Use		AZ 91C		AM 100A		QE 22A
Cross-Reference Specification	Code No.	21465		21327		21535
	AMS	4437		4455		4418
	SAE	504		502		---
	ASTM	B-403		B-403		B-403
	Federal QQ	M-55a		M-55		---
	MIL	---		---		---
Chemical Composition* per cent (Balance Magnesium)	Al	8.1-9.3		9.3-10.7		---
	Mn	0.13		0.10 min.		---
	Si	0.30		0.30		---
	Zn	0.40-1.0		0.30		---
	Zr	---		---		0.40-1.0
	Th	---		---		---
	Rare Earths	---		---		1.8-2.5
	Cu	0.10		0.10		0.10
	Ni	0.01		0.01		0.01
	Ag	---		---		2.0-3.0[(1)]
Temper		F	T4	F	T6	T6
Mechanical Properties**	U.T.S., psi	18,000	34,000	20,000	34,000	35,000
	Y.S., 0.2%, psi	10,000	10,000	10,000	15,000	25,000
	% E--1"	Not req.	7	Not req.	2	2
Characteristics		Best general purpose magnesium alloy-can be heat treated and welded; used for pressure-tight castings; service temperature up to 300F		For pressure tight castings with good mechanical properties; corrosion resistance decreases with increasing Cu, Ni and Fe contents; can be brazed for assemblies		High yield strength with excellent creep resistance and fatigue strength at service temperatures up to 500F; used in aircraft, missiles and space vehicles

*   Figures are maximum values unless stated as a range.
**   Figures are minimum values unless stated as a range.
(1)   AMS 4418 requires 2.0-3.0 Ag.

OTHER MAGNESIUM ALLOYS FOR SPECIAL PURPOSES

Designation of other Alloys for General Use		AZ 92A	EZ 33A	ZK 61A
Cross-Reference Specification	Code No.	21455	21531	21520
	AMS	4453		
		4484	4442	4444
	SAE	500	506	513
	ASTM	B-403	B-403	B-403
	Federal QQ	M-56b	M-56b	M-56b
	MIL	C-19163	---	---
Chemical Composition* per cent (Balance Magnesium)	Al	8.3-9.7	---	---
	Mn	0.10 min.	---	---
	Si	0.30	---	---
	Th	1.6-2.4	2.0-3.1	5.5-6.5
	Zr	---	0.50-1.0	0.6-1.0
	Zn	---	---	---
	Rare Earths	---	2.5-4.0	---
	Cu	0.10	0.10	0.10
	Ni	0.01	0.01	0.01
	Ag	---	---	---

TABLE 8 - NICKEL BASE ALLOYS
Typical Chemical Range Percentages

Trade Name	C	Mn	P	S	Si	Ni	Cr	Mo	Cu	Fe	Co	V	W	Other
Alloy B	.12	1.0	.030	.030	1.00	Bal.	1.00	26.00 30.00		4.00 6.00	2.50	.20 .60		
Alloy C (CW-6M)	.07	1.0	.04	.03	1.00	Bal.	17.0 20.0	17.0 20.0		3.00				
Alloy C (CW-12MW)	.12	1.0	.03	.03	1.0	+ Co Bal.	15.5 17.5	16.0 18.0		4.5 7.0	2.5	.20 .40	3.75 5.25	
Alloy X	.20	1.0	.04	.03	1.00	Bal.	20.50 23.00	8.00 10.00		17.00 20.00	.50 2.50		.20 1.00	
Invar	.12	.20 .50	.04	.03	.40	35.0 37.0				Bal.				
In 600 ①	.15	1.0	.03	.015	.50	72 min.	14.0 17.0		.50	6.0 10.0				
In 625 ②	.10	.50	.015	.015	.50	Bal.	20.0 23.0	8.0 10.0	.30	5.0	1.0			0.4Ti, 0.41Al, 3.15-4.15 Cb + Ta
Monel 410	.35	1.5			2.0	62.0 68.0			26.0 33.0	2.5	*			.50 Al
S Monel	.25	1.5			3.5 4.5	60.0			27.0 31.0	2.5	*			.50 Al
RH Monel	.40 .70	1.5			2.3 3.0	Bal.			29.0 34.0	.5	1.0			.50 Al
E Monel	.30	1.5			1.0 2.0	60.0 min.			26.0 33.0	3.5	*			.50 Al 1.0- 3.0Cb + Ta
M-35 Monel	.35	1.5	.03	.03	1.25	Bal.			26.0 33.0	3.5				0.5 Max Cb
47-50	.05	.60	.02	.02	.60	47.0 50.0				Bal.				

Cobalt included in % Nickel. Maximum unless range is shown.
① Also see International Nickel's Inconel 600.
② Also see International Nickel's Inconel 625.

TABLE 9: PROPERTIES OF SEPARATELY CAST TEST BARS OF NICKEL BASE ALLOYS

Alloy	Condition	Tensile Strength		0.2% Yield Strength		% Elongation Range (in 2.5cm)	Hardness Range
		English psi	Metric MPa	English psi	Metric MPa		
Alloy B	Annealed	75-85,000	517-586	50-60,000	345-414	8-12	90-100 Rb
Alloy	As Cast	80-95,000	552-655	45-55,000	310-379	8-12	90-100 Rb
C	Annealed	75-95,000	517-655	45-55,000	310-379	8-12	90 Rb - 25 Rc
Alloy	A.C.24°C	63-70,000	434-483	41-45,000	283-310	10-15	85-96 Rb
X	A.C.816°C	35-45,000	241-310			12-20	
Invar	As Cast	50-60,000	345-414	25-30,000	172-207	30-40	50-60 Rb
In 600①	As Cast	65-75,000	448-517	35-40,000	241-276	10-20	80-90 Rb
In 625②	Annealed	80-100,000	552-758	40-55,000	276-379	15-30	10-20 Rc
Monel 410	As Cast	65-75,000	448-517	32-38,000	221-262	25-35	65-75 Rb
S Monel	Annealed	100-110,000	690-758	55-65,000	379-448	5-10	20-28 Rc
Monel	Hardened	120-140,000	827-965	85-100,000	586-690	0	32-38 Rc
RH Monel	As Cast	100-110,000	690-758	60-80,000	414-552	10-20	20-30 Rc
Monel E	As Cast	65-80,000	448-552	33-40,000	227-276	25-35	67-78 Rb
M-35 Monel	As Cast	65-80,000	448-552	25-35,000	172-241	25-40	65-85 Rb
47-50	Annealed	60-70,000	414-483	20-30,000	138-207	25-35	55-60 Rb

NOTE: The mechanical property values shown in Table 9 are for information only. They do not necessarily apply to casting. Any
• requirements for mechanical properties are beyond this standard and must be negotiated with the foundry.
①   Also see International Nickel's Inconel 600.
②   Also see International Nickel's Inconel 625.

TABLE 10: IRON, CARBON AND LOW ALLOY STEELS
Typical Chemical Range Percentages (%)*

ALLOY	C	Mn	Si	Ni	Cr	Mo	P	S	Other
1.2% Silicon Iron	.04	---	.90 / 1.3	---	---	---	.01	.03	.15 C + P + S + Mn
2.5% Silicon Iron	.04	---	2.3 / 2.7	---	---	---	.01	.03	.15 C + P + S + Mn
IC 1010	.05 / .15	.30 / .60	.40 / .80	---	---	---	.04	.04	
IC 1020	.15 / .25	.20 / .60	.20 / 60	---	---	---	.04	.045	
IC 1030	.25 / .35	.70 / 1.0	.20 / .60	---	---	---	.04	.045	
IC 1040	.35 / .45	.70 / 1.0	.20 / 1.0	---	---	---	.04	.045	
IC 1050	.45 / .55	.7 / 1.0	.20 / 1.0	---	---	---	.04	.045	
IC 1060	.55 / .65	.60 / .90	.20 / 1.0	---	---	---	.04	.045	
IC 1090	.85 / .98	.60 / .90	.20 / 1.0	---	---	---	.04	.045	
IC 2345	.40 / .50	.70 / .90	.20 / .80	3.25 / 3.75	---	---	.04	.04	
IC 3120	.15 / .25	.60 / .80	.20 / .80	1.1 / 1.4	.55 / .75	---	.04	.04	
IC 4130	.25 / .35	.40 / .70	.20 / .80	---	.80 / 1.10	.15 / .25	.04	.04	
IC 4140	.35 / .45	.70 / 1.0	.20 / .80	---	.80 / 1.10	.15 / .25	.04	.04	
IC 4150	.45 / .55	.75 / 1.0	.20 / .80	---	.80 / 1.10	.15 / .25	.04	.04	
IC 4330	.28 / .36	.60 / 1.0	.20 / .80	1.65 / 2.0	.65 / 1.0	.30 / .45	.04	.04	
IC 4340	.36 / .44	.60 / .90	.20 / .80	1.65 / 2.0	.70 / .90	.20 / .30	.025	.025	
IC 4620	.15 / .25	.40 / .70	.20 / .80	1.65 / 2.0	---	.20 / .30	.04	.045	
IC 6120	.15 / .25	.70 / .90	.20 / .80	---	.70 / 1.0	---	.04	.04	
IC 6150	.45 / .55	.65 / .95	.20 / .80	---	.80 / 1.10	---	.04	.045	V .15 Min.
IC 8620	.15 / .25	.65 / .95	.20 / .80	.40 / .70	.40 / .70	.15 / .25	.04	.045	
IC 8630	.25 / .35	.65 / .95	.20 / .80	.40 / .70	.40 / .70	.15 / .25	.04	.045	
IC 8640	.35 / .45	.70 / 1.05	.20 / .80	.40 / .70	.40 / .60	.15 / .25	.04	.04	
IC 8665	.60 / .70	.70 / 1.05	.20 / .80	.40 / .70	.40 / .60	.15 / .25	.04	.04	
IC 8730	.25 / .35	.70 / .90	.20 / .80	.40 / .70	.40 / .60	.15 / .30	.04	.04	
IC 8740	.35 / .45	.75 / 1.0	.20 / .80	.40 / .70	.40 / .60	.20 / .30	.04 / .04	.04 / .04	
IC 52100	.95 / 1.10	.25 / .55	.20 / .80	---	1.30 / 1.60	---	.04	.045	
IC 1722AS	.27 / .34	.45 / .65	.55 / .75	---	1.0 / 1.5	.40 / .60	.04	.04	
Ductile Iron Ferritic	3.2 / 4.0	.25 / .65	2.0 / 2.6	---	---	---	.03	.03	.03-.08 Mg
Ductile Iron Pearlitic	3.3 / 4.0	.50 / .70	2.1 / 2.5	1.5 / 2.0	---	.30 / .50	.03	.03	.03-.08 Mg

* *Where not specified Ni, Cr, Mo and Cu shall not exceed .25% ea. They will not be analyzed for unless by previous agreement between purchaser and foundry.*

TABLE 11: PROPERTIES OF SEPARATELY CAST TEST BARS OF IRON, CARBON AND LOW ALLOY STEELS

Alloy	Condition	Tensile Strength		0.2% Yield Strength		% Elongation Range (in 2.54 cm)	Hardness Range or Max
		English psi	Metric MPa	English psi	Metric MPa		
1.2% Si		50-60,000	345-414	37-43,000	255-296	30-35	55 Rb
2.5% Si						0	85 Rb
IC 1010	Annealed	50-60,000	345-414	30-35,000	207-241	30-35	50-55 Rb
IC 1020	Annealed	60-70,000	414-483	40-45,000	276-310	25-40	80 Rb
IC 1025	Annealed	63-73,000	434-503	42-47,000	290-324	25-35	80 Rb
IC 1030	Annealed	65-75,000	448-517	45-50,000	310-345	20-30	75 Rb
	Hardened	85-150,000	586-1034	60-150,000	414-1034	0-15	20-50 Rc
IC 1035	Annealed	70-80,000	483-552	45-55,000	310-379	20-30	80 Rb
	Hardened	90-150,000	621-1034	85-150,000	586-1034	0-15	25-52 Rc
IC 1045	Annealed	80-90,000	552-621	50-60,000	345-414	20-25	100 Rb
	Hardened	100-180,000	690-1241	90-180,000	621-1241	0-10	25-57 Rc
IC 1050	Annealed	90-110,000	621-758	50-65,000	345-448	20-25	100 Rb
	Hardened	125-180,000	862-1241	100-180,000	690-1241	0-10	30-60 Rc
IC 1060	Annealed	100-120,000	690-827	55-70,000	379-483	12-20	25 Rc
	Hardened	120-200,000	827-1379	100-180,000	690-1241	0-5	30-60 Rc
IC 1090	Annealed	110-150,000	758-1034	70-80,000	483-552	5-10	30 Rc
	Hardened	130-180,000	896-1241	130-180,000	876-1241	0-3	37-50 Rc
IC 2345	Annealed	---	---	---	---	---	100 Rb
	Hardened	130-200,000	896-1394	110-180,000	758-1241	5-10	30-58 Rc
IC 3120	Annealed	---	---	---	---	---	100 Rb
IC 4130	Annealed	---	---	---	---	---	100 Rb
	Hardened	130-170,000	896-1172	100-130,000	690-896	5-20	23-49 Rc
IC 4140	Annealed	---	---	---	---	---	100 Rb
	Hardened	130-200,000	876-1394	100-155,000	690-1069	5-20	29-57 Rc
IC 4150	Annealed	---	---	---	---	---	100 Rb
	Hardened	140-200,000	965-1394	120-180,000	827-1241	5-10	25-58 Rc
IC 4330	Annealed	---	---	---	---	---	20 Rc
	Hardened	130-190,000	876-1310	100-175,000	690-1207	5-20	25-48 Rc
IC 4340	Annealed	---	---	---	---	---	20 Rc
	Hardened	130-200,000	876-1394	100-180,000	690-1241	5-20	20-55 Rc
IC 4620	Annealed	---	---	---	---	---	100 Rb
	Hardened	110-150,000	758-1034	90-130,000	621-896	10-20	20-32 Rc
IC 6120	Annealed	---	---	---	---	---	100 Rb
IC 6150	Annealed	---	---	---	---	---	100 Rb
	Hardened	140-200,000	965-1394	120-180,000	827-1241	5-10	30-60 Rc
IC 8620	Annealed	---	---	---	---	---	100 Rb
	Hardened	100-130,000	690-896	80-110,000	552-758	10-20	20-45 Rc
IC 8630	Annealed	---	---	---	---	---	100 Rb
	Hardened	120-170,000	827-1172	100-130,000	690-896	7-20	25-50 Rc
IC 8640	Annealed	---	---	---	---	---	20 Rc
	Hardened	130-200,000	876-1394	100-180-000	690-1241	5-20	30-60 Rc
IC 8665	Annealed	---	---	---	---	---	25 Rc
	Hardened	170-220,000	1172-1517	140-200,000	965-1394	0-10	---
IC 8730	Annealed	---	---	---	---	---	100 Rb
	Hardened	120-170,000	827-1172	110-150,000	758-1034	7-20	---
IC 8740	Annealed	---	---	---	---	---	100 Rb
	Hardened	140-200,000	965-1394	120-180,000	827-1241	5-10	30-60 Rc
IC 52100	Annealed	---	---	---	---	---	25 Rc
	Hardened	180-230,000	1241-1585	140-180,000	965-1241	1-7	30-65 Rc
IC 1722AS	Annealed	---	---	---	---	---	25 Rc
	Hardened	130-170,000	876-1172	100-140,000	690-1241	6-12	25-48 Rc
Ductile Iron Ferritic	Annealed	60-80,000	414-552	40-50,000	276-345	18-24	(143-200 BHN)
Ductile Iron Pearlitic	Normalized	100-120,000	690-830	70-80,000	483-552	3-10	(243-303 BHN)

NOTE: The above mechanical property values are for information only. They do not necessarily apply to casting. Any requirements for mechanical properties are beyond this standard and must be negotiated with the foundry.

TABLE 12: HARDENABLE STAINLESS STEELS
Typical Chemical Range Percentages

Trade Name (AISI Equivalent)	C	Mn	Si	P	S	Ni	Cr	Mo	Cu	Other
CA-15 (410)	.05 .15	1.00	1.50	.04	.04	1.00	11.5 14.0	.50	---	
IC 416 (416)	.15	1.25	1.50	.05	.15 .35	.50	11.50 14.0	.50	.50	.10-.30Se alternate for S, .50Zr
CA-40 (420)	.20 .40	1.00	1.50	.04	.04	1.0	11.5 14.0	.5	---	
IC 431 (431)	.08 .15	1.00	1.00	.04	.04	1.50 2.20	15.0 17.0	---	---	0.22 C + N; .03-.12N*
IC 440A (440A)	.60 .75	1.00	1.00	.04	.03	---	16.0 18.0	.75	---	
IC 440C (440C)	.95 1.20	1.00	1.00	.04	.03	.75	16.0 18.0	.35 .75	---	
IC 440F (440F)	.95 1.2	1.00	1.00	.04	.15 .35	.50	16.0 18.0	.75	.50	.10-.30Se alternate for S
Greek Ascoloy	.15 .20	1.00	1.00	.04	.03	1.80 2.20	12.00 14.00	.50	.50	2.50-3.5W
IC 17-4	.06	.70	.50 1.00	.04	.03	3.60 4.60	15.50 16.70	---	2.80 3.50	.15-.40 Cb + Ta, .05N
AM-355	.08 .15	.40 1.10	.75	.04	.03	3.50 4.50	14.50 15.50	2.00 2.60		.05-.13N; .15-.25C + N
CA-6NM	.06	1.00	1.00	.04	.03	3.5 4.5	11.5 14.0	.40 1.0	---	
IC 15-5	.05	.60	.50 1.00	.025	.025	4.20 5.00	14.00 15.50	---	2.50 3.20	.15-.30 Cb + Ta, .05N
CD-4MCu	.04	1.00	1.00	.04	.04	4.75 6.00	24.5 26.5	1.75 2.25	2.75 3.25	

* Nitrogen analysis is not routinely measured for commercial castings. To assure hardenability, nitrogen must be in range and its analysis should be requested by the purchaser.

TABLE 13: PROPERTIES OF SEPARATELY CAST TEST BARS
OF HARDENABLE, STAINLESS STEELS

Alloy	Condition	Tensile Strength		0.2% Yield Streng		% Elongation Range (in 2.5 cm)	Hardness Range or Max
		English psi	Metric MPa	English psi	Metric MPa		
CA-15	Annealed	---	---	---	---	---	100 Rb
	Hardened	95-200,000	655-1394	75-160,000	517-1103	5-12	94 Rb-45 Rc
IC 416	Annealed	---	---	---	---	---	100 Rb
	Hardened	95-200,000	655-1394	75-160,000	517-1103	3-8	94 Rb-45 Rc
CA-40	Annealed	---	---	---	---	---	25 Rc
	Hardened	200-225,000	1394-1551	130-210,000	896-1448	0-5	30-52 Rc
IC 431	Annealed	---	---	---	---	---	30 Rc
	Hardened	110-160,000	759-1103	75-105,000	517-724	5-20	20-40 Rc
IC 440A	Annealed	---	---	---	---	---	30 Rc
	Hardened	---	---	---	---	---	35-56 Rc
IC 440C	Annealed	---	---	---	---	---	35 Rc
	Hardened	---	---	---	---	---	40-60 Rc
IC 440F	Annealed	---	---	---	---	---	35 Rc
	Hardened	---	---	---	---	---	40-60 Rc
Greek Ascoloy	Annealed	---	---	---	---	---	36 Rc
IC 17-4	Annealed	---	---	---	---	---	36 Rc
	Hardened	150-190,000	1034-1310	140-160,000	965-1103	6-20	34-44 Rc
AM-355	Hardened	200-220,000	1394-1517	150-165,000	1034-1138	6-12	---
IC 15-5	Hardened	135-170.000	931-1172	110-145,000	759-1000	5-15	26-38 Rc
CD-4MCu	Annealed	100-115,000	690-793	75-85,000	517-586	20-30	94-100 Rb
	Hardened*	135-145,000	931-1000	100-120,000	690-827	10-25	28-32 Rc

NOTE: The above mechanical property values are for information only. They do not necessarily apply to casting.
Any requirements for mechanical properties are beyond this standard and must be negotiated with the foundry.
* Not generally sold in hardened condition.

TABLE 14: AUSTENITIC STAINLESS STEELS
Typical Chemical Range Percentages

Trade Name (AISI Equivalent)	C	Mn	Si	P	S	Ni	C	Mo	Cu	Other
CF-20 (302)	.20	1.50	2.00	.04	.04	8.0 11.0	18.0 21.0	---	---	
CF-16F (303)	.16	1.50	2.00	.04	---	9.0 12.0	18.0 21.0	---	---	Either: .20-.35Se, 1.50 Mo or .40-.80Mo, .20-.40S
CF-8 (304)	.08	1.50	2.00	.04	.04	8.0 11.0	18.0 21.0	---	---	
CF-3 (304L)	.03	1.50	2.00	.04	.04	8.0 12.0	17.0 21.0	---	---	
CH-20 (309)	.20	1.50	2.00	.04	.04	12.0 15.0	22.0 26.0	---	---	
CK-20 (310)	.20	2.00	2.00	.04	.04	19.0 22.0	23.0 27.0	---	---	
CF-8M (316)	.08	1.50	2.00	.04	.04	9.0 12.0	18.0 21.0	2.0 3.0		
CF-3M (316L)	.03	1.50	1.50	.04	.04	9.0 13.0	17.0 21.0	2.0 3.0	---	
IC 316F (316F)	.08	1.50	2.0	.04	.04	9.0 12.0	18.0 21.0	2.0 3.0		.20-.35Se or .20-.40S
IC 321 (321)*	.08	2.00	1.00	.04	.03	9.0 12.0	17.0 19.0	---		Ti = (5xC) (min)
CF-8C (347)	.08	1.50	2.00	.04	.04	9.0 12.0	18.0 21.0	---	---	Cb = (8xC) (min) - 1.0Cb (max)
CN-7M	.07	1.50	1.50	.04	.04	27.5 30.5	19.0 22.0	2.0 3.0	3.0 4.0	
HK	.20 .60	2.00	2.00	.04	.04	18.0 22.0	24.0 28.0	.50	---	

* CF-8C is recommended in lieu of IC-321 for castability

TABLE 15: PROPERTIES OF SEPARATELY CAST TEST BARS
OF AUSTENITIC STAINLESS STEELS

Alloy	Condition	Tensile Strength		0.2% Yield Strength		% Elongation Range (in 2.5 cm)	Hardness R_b Max.
		English psi	Metric MPa	English psi	Metric MPa		
CF-20	Annealed	65-75,000	448-517	30-35,000	207-241	35-60	90
CF-3, CF-8	Annealed	70-85,000	483-586	40-50,000	276-345	35-50	90
CH-20	Annealed	70-80,000	483-552	30-40,000	207-276	30-45	90
CK-20	Annealed	60-75,000	414-517	30-40,000	207-276	35-45	90
CF-3M, -8M, IC 316F	Annealed	70-85,000	483-586	40-50,000	276-345	35-50	90
CF-16F	Annealed	65-75,000	418-517	30-35,000	207-241	35-45	90
CF-8C	Annealed	70-85,000	483-586	32-36,000	221-248	30-40	90
CN-7M	Annealed	65-75,000	418-517	25-35,000	172-241	35-45	90
IC 321	Annealed	65-75,000	418-517	30-40,000	207-276	35-45	90
HK	Annealed	65-75,000	418-517	35-45,000	241-310	10-20	100

TABLE 16: TOOL STEELS
Typical Chemical Range Percentages

Alloy	Carbon	Manganese	Silicon	Chromium	Molybdenum	Tungsten	Vanadium	Other
CA-2	.95 1.05	.75	1.50	4.75 5.50	.90 1.40	---	.20-.50 Optional	
CA-6	.65 .75	1.80 2.20	1.00	.80 1.20	.80 1.30	---	---	
CD-2	1.40 1.60	1.00	1.50	11.0 13.00	.70 1.20	---	.40-1.00 Optional	
CD-3	2.10 2.30	.75	1.00	11.5 13.0	.40	---	---	
CD-6	2.10 2.35	.75	.80 1.20	11.5 13.0	.40	.80 1.20	---	
CD-7	2.15 2.45	.75	1.00	11.5 13.0	.80 1.20	---	3.50-4.50	
CH-11	.30 .40	.75	.95 1.15	4.6 5.4	1.20 1.60	---	.30-.50	
CH-12	.30 .40	.75	1.50	4.75 5.75	1.25 1.75	1.00 1.70	.20-.50	
CH-13	.30 .42	.75	1.50	4.75 5.75	1.25 1.75	---	.75-1.20	
CL-6	.65 .75	.75	1.00	.80 1.0	---	---	---	1.50-1.90 Ni
C1-M-2	.95 1.05	.75	1.00	3.75 4.50	4.50 5.50	5.50 6.75	1.75 2.20	.25Ni
CM-2	.78 .88	.75	1.00	3.75 4.50	4.50 5.50	5.50 6.75	1.25 2.20	.25Ni
CM-4	1.25 1.35	.75	1.00	3.75 4.50	4.50 5.50	5.20 6.20	3.60-4.40	
CM-42	1.00 1.20	.75	1.00	3.50 4.25	9.00 10.00	1.25 1.75	.95 1.35	7.50-8.50 Co
CM-43	1.15 1.35	.75	1.00	3.50 4.25	8.25 9.25	1.50 2.00	1.50 2.00	7.75-8.75 Co
CO-1	.85 1.00	1.00 1.30	1.50	.40 1.00	---	.40 .60	.30	
CO-2	.85 .95	1.50 1.80	1.00	.40	.30	---	.30	
CO-7	1.10 1.20	.75	1.00	.50 .70	---	1.65 1.85	.15 .25	
CS-1	.45 .55	.75	1.00	1.35 1.65	---	2.35 2.65	---	
CS-2	.45 .55	.75	.90 1.20	---	.40 .60	---	.30	
CS-4	.50 .60	.70 .90	1.80 2.20	.30	---	---	.30	
CS-5	.50 .65	.60 1.00	1.75 2.25	.35	.20 .80	---	.35	
CS-7	.50 .60	.50 .80	1.00	3.00 3.50	1.20 1.60	---	---	
CT-1	.65 .75	.75	1.00	3.75 4.50	---	17.25 18.75	.90 1.30	
CT-2	.80 .90	.75	1.00	3.75 4.50	1.00	17.50 19.00	1.80 2.40	
CT-6	.75 .85	.75	1.00	4.00 4.75	.70 1.00	18.50 21.25	1.50 2.10	10.00- 13.70 Co

* *Max., unless a range is shown .025% S and .025% P max. all grades.*

TABLE 17: HARDNESS VALUES OF CASTINGS AND
SEPARATELY CAST TEST BARS OF TOOL STEELS

ALLOY	HARDNESS		
	Annealed with Slow Cool Max.	Cycle Anneal Max.	Hardened Range (Rc)
CA-2	20 Rc	27 Rc	47-60
CA-6	100 Rb		48-59
CD-2		35 Rc	50-59
CD-3		35 Rc	47-61
CD-6	100 Rb		50-63
CD-7	24 Rc		50-63
CH-11	100 Rb		46-55
CH-12	100 Rb		50-53
CH-13	100 Rb		45-53
CL-6	95 Rb		39-60
C1-M-2		30 Rc	61-63
CM-2		30 Rc	61-63
CM-4	30 Rc		62-64
CM-42		35 Rc	60-64
CM-43	27 Rc		61-64
CO-1		100 Rb	45-61
CO-2		100 Rb	38-60
CO-7	95 Rb		35-64
CS-1		100 Rb	44-57
CS-2		100 Rb	44-55
CS-4	100 Rb		42-53
CS-5	100 Rb		37-59
CS-7		100 Rb	35-57
CT-1	100 Rb		60-66
CT-2	100 Rb		60-66
CT-6	30 Rc		60-64

Alloys for Vacuum Investment Casting

Developments in processing techniques and in applications of investment castings produced from vacuum melted alloys have been rapid in the past several years. Many vacuum melted metals, particularly high temperature high stress compositions, are well suited to production of investment castings with close dimensional tolerance and excellent detail.

Although many alloys may be prepared and cast under vacuum conditions, major benefits have been found in nickel and cobalt base alloys containing substantial quantities of aluminum and titanium. The use of this process permits alloy preparation relatively free from inclusions, greatly reduces formation of aluminum oxide and the nitrides and oxides of titanium, thus permitting the titanium and aluminum to function more efficiently in a strengthening mechanism. In general, vacuum melting has improved structure-sensitive properties such as fatigue strength, ductility, stress rupture properties and impact strength; its effect on tensile strength is negligible.

The vacuum investment casting field comprises four phases:
1. Purpose of vacuum melting
2. Equipment
3. Processing
4. Alloys and properties

Purpose of vacuum melting

Vacuum melting, refining and casting can have these beneficial effects on the product being prepared:

A. Control of gas content
   Melt-atmosphere reactions
   Deoxidation practice
   Gas evolution on solidification
B. Control of minor elements (normally non-gaseous)
   Distillation
   Refractory (and oxide) reduction
   Control of volatile impurities.
C. Interactions and controlled additions
   Interactions of A and B
   Addition of reactive elements (i.e. Ti and Al)

An important factor which must not be overlooked is vapor pressure of the various elements within the melt composition. Elements with high vapor pressures will evaporate from the melt during the melt cycle, which may create a problem in control of final alloy composition. The amount of element loss is a function of pressure, temperature and time.

Figure 6:1 shows a vacuum melting unit with a vertical mold lock configuration.

Figure 6:1

TABLE 18: VACUUM CAST ALLOYS

VACUUM CAST ALLOYS

NICKEL BASE ALLOYS

Alloy Designation		IN-100	Mar-M 247	IN-718	IN-713C	IN-713LC	B-1900
Chemical Composition	C	0.15-0.2	0.13-0.17	0.08 Max.	0.08-0.20	0.05	0.10
per cent:	Mn	0.10	-	0.35 Max.	0.25	-	-
	Si	0.15	-	0.35 Max.	0.50	-	-
	Cr	8.0-11.0	8.0-8.8	17.00 - 21.00	12-14	12.0	8.0
	Ni	Bal.	Bal.	50.00 - 55.00	Bal.	Bal.	Bal.
	Co	13.0-17.0	9.0-11.0	1.00 Max.	-	-	10.0
	Fe	1.0	-	Bal.	2.5	-	-
	Mo	2.0-4.0	0.5-0.8	2.80 - 3.30	3.8-5.2	4.5	6.0
	W	-	9.5-10.5	-	-	-	-
	Al	5.0-6.0	5.3-5.7	0.40 - 0.80	5.5-6.5	5.9	6.0
	Ti	4.5-5.0	0.9-1.2	0.65 - 1.15	0.5-1.0	0.6	1.0
	B	0.01-0.02	0.01-0.02	0.0006 Max.	0.005-0.015	0.010	0.015
	Zr	0.03-0.09	0.03-0.08	-	0.05-0.15	0.10	0.10
	Cu	-	-	0.30 Max.	0.50	-	-
	Cb+Ta	-	2.8-3.3 Ta	0.10 Max.	Cb 1.8-2.8	Cb 2.0	4.0 Ta
	V	0.70-1.2	-	-	-	-	-
	S	0.015	-	0.015 Max.	0.015	-	-
	Other	-	Hf 1.2-1.6	1.75 Max. (Ti & Al)	-	-	-

Mechanical Properties Heat Treated Condition	IN-100	Mar-M 247	IN-718	IN-713C	IN-713LC	B-1900
Test Temperature °F	RT 1400 1800	RT 1400 1800	RT 1200	RT 1400 1800	RT 1400 1800	RT 1400 1800
UTS max. psi x1000	147 155 82	140 150 80	150 121	123 136 68	130 138 68	141 138 80
YS (0.2% offset) psi x1000	123 125 54	118 120 55	130 110	107 108 44	109 110 41	120 110 41
Elongation %	9 6.5 6.0	7 - -	7 10	8 6 20	15 11 22	8 4 7
Reduction of Area %			20 26	- - -	- - -	- - -
100 hr Stress Rupture psi x1000	- 91 25	- 100 27	- -	- 83 21	- 80 20	- 56 25
1000 hr Stress Rupture psi x1000-	- 75 15	- - 18	- -	- 65 13	- 65 -	- - -

Applications & Notes	IN-100	Mar-M 247	IN-718	IN-713C	IN-713LC	B-1900
	High Strength alloy for turbine blades.	Mfb properties High strength alloy for blades & wheels. ¹HT: 1600°F 16 hr air cool	Airframe & jet engine hardware and structural components with service temp. up to 1200°F.	Turbine blades & vanes and other applications requiring high strength & creep resistance to 1700°F.	See IN-713C.	High Strength blade material

Check with your Investment Casting Foundry to determine the best vacuum cast alloy for your application.

TABLE 18: VACUUM CAST ALLOYS (Cont'd)

VACUUM CAST ALLOYS

		NICKEL BASE ALLOYS Cont'd.		COBALT BASE ALLOYS		HIGH STRENGTH STEELS
Alloy Designation		GMR235 D	UDIMET 500	MAR-M 50 9	WI-52	A-286
Chemical Composition per cent:	C	0.15	0.13	0.55-0.65	0.45	0.08
	Mn	-	0.20	0.1 max.	0.25	2.0
	Si	-	0.30	0.4 max.	0.25	1.0
	Cr	15.15	16.20	22.5-24.5	21.0	13.5-16.0
	Ni	Bal	Bal	9.0-11.0	-	24.8-27.0
	Co	-	16.0-20.0	Bal.	Bal.	-
	Fe	4.5	2.0	1.5 max.	2.0	Bal.
	Mo	5.0	3.0-5.0	-	-	1.0-1.5
	W	-	-	6.5-7.5	11.0	-
	Al	3.5	2.5-3.0	-	-	0.35
	Ti	2.5	2.5-3.25	0.15-0.35	-	1.9-2.3
	B	0.05	0.007	-	-	0.003-0.01
	Zr	-	0.05	0.3-0.6	-	-
	Cu	-	-	-	-	-
	Cb+Ta	-	-	3.0-4.0 Ta	2.0 Cb	-
	V	-	-	-	-	0.1-0.5
	S	-	-	0.015 max	-	-
	Other	-	-	-	-	-

Mechanical Properties Heat Treated Condition	As Cast		HT[3]			As Cast			As-Cast			HT[4]		
Test Temperature °F	RT	1400	1800	RT	1400	1800	RT	1400	1800	RT	1400	1800	RT	1400	1800
UTS max. psi x1000	112	115	-	135	124	19	114	83	31	109	88	40	106	-	-
YS (0.2% offset) psi x1000	103	86	-	118	102	-	83	53	26	85	50	28	85.4	-	-
Elongation %	3.5	3.0	-	13	9	50	4	10	26	5	9	20	10.5	-	-
Reduction of Area %	-	-	-	-	-	-	-	-	-	-	-	-	17.0	-	-
100 hr Stress Rupture psi x1000	-	77	17	-	65	13	-	50	15	-	-	13	-	-	-
1000 hr Stress Rupture psi x1000	-	60	12	-	50	-	-	38	11.5	-	-	10	-	-	-

| Applications & Notes | Turbine rotating parts. | Turbine blades & other parts requiring high strength and corrosion resistance to 1700°F. 3. HT: 2100F/4/AC 1975F/4/AC 1400F/16/AC | High strength high hot corrosion resistance for vanes etc. | High strength + hot cor- rosion resistance for stator castings. | Structural parts good strength and corrosion resistance to 1200°F. 4HT: 2000F/2/00 1800F/1/AC 1325F/16/AC. |

Check with your Investment Casting Foundry to determine the best vacuum cast alloy for your application.

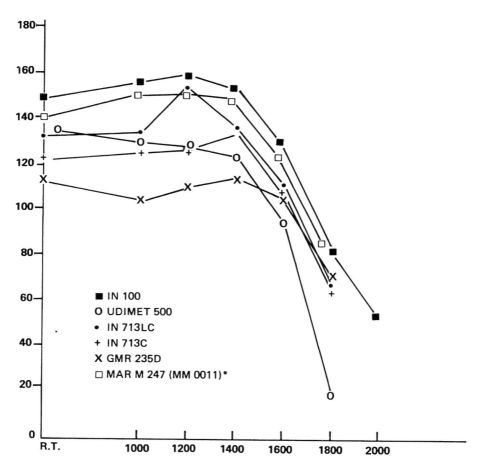

* MfB Properties

ULTIMATE TENSILE STRENGTH
CAST HIGH TEMPERATURE ALLOYS

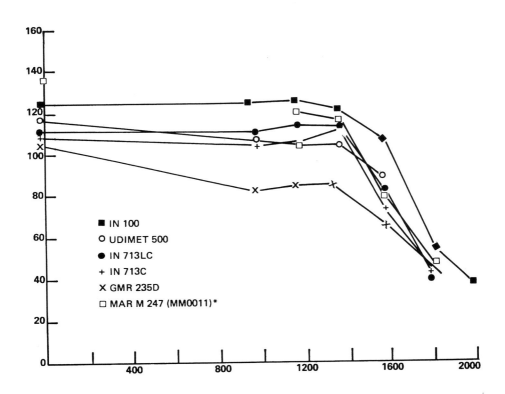

*MfB. Properties

TEMPERATURE °F
1000 HOUR RUPTURE STRENGTH
CAST HIGH TEMPERATURE ALLOYS

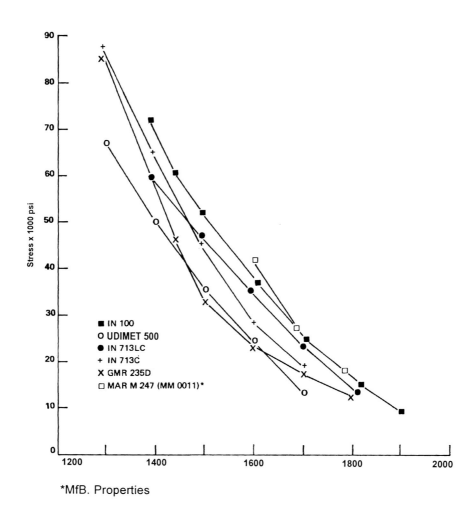

*MfB. Properties

TEMPERATURE °F
1000 HOUR RUPTURE STRENGTH
CAST HIGH TEMPERATURE ALLOYS

Titanium

The investment casting of titanium and its alloys has grown in importance and value over the years. A wide range of components for aircraft applications are readily produced. The strength-to-weight ratio of titanium alloys was often a critical factor in their selection for these applications. Titanium alloys have to be vacuum cast and the process costs involved often limited the applications of these alloys. Process improvements in vacuum melting and ceramic mold systems has permitted quite spectacular growth in the production of titanium castings for aerospace, medical and sports equipment applications. Of particular note is the phenomenal growth in the production of titanium golf club heads.

Figure 6:2 shows a cast titanium auxiliary power unit duct.

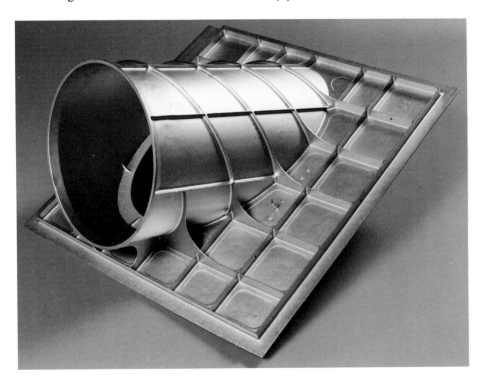

Figure 6:2
BOEING
777 APU DUCT
Titanium Investment Castings

There are two major titanium alloy groups which are investment cast for aerospace type applications. The table lists typical chemistries and properties for Ti-6-4 and Ti-6-24.2 alloys

TABLE 19: TITANIUM ALLOYS

Alloy	Ti - 6A1 - 4V		Ti - 6AI - 2SM - 4ZR - 2 MO	
Chemistry	Min.	Max.	Min.	Max.
Al	5.50	6.75	5.50	6.50
Vanadium	3.50	4.50	-	-
Tin	-	-	1.50	2.50
Molybdenum	-	-	1.50	2.50
Zirconium	-	-	3.60	4.40
Iron	-	0.30	-	0.35
Carbon	-	0.08	-	0.08
Hydrogen	-	0.015	-	0.015
Oxygen	-	0.20	-	0.012
Yttrium	-	-	-	-
Others Total	-	0.40	-	0.40
Ti	BAL.		BAL.	
HEAT TREATMENT				
HIP	1650 or 1750°F - 15 KSI - 2 Hrs.		1650°F or 1750°F - 15 KSI - 2 Hrs.	
HT Condition	1550°F 2 Hrs.		1750°F 1 Hr. + 1100°F 8 Hrs.	
TYPICAL MECHANICAL PROPERTIES				
TEST BAR TYPE	CAST TO SIZE		CAST TO SIZE	
TEST TEMP.	70°F		70°F	1000°F
UTS KSI	136		145	95
0.2% Yield KSI	120		126	72
ELONGATION %	12		11	18
ROA %	22		20	45
APPLICATIONS	Airframe & Jet Engine Components Service Temp. up to 600°F		Airframe & Jet Engine Components Service temp. up to 1000°F	

Glossary

Acid Etch—See Etching.

Air Bubbles—Entrapped air globules on pattern during dip coating or investing that result in positive metal on casting.

Air-Dried—Mold dried in air, without application of heat.

Air-Dried Strength—Tenacity (compressive, shear, tensile, or transverse) of a refractory mixture after being air-dried at room temperature.

Air Setting— The characteristics of some materials, such as refractory cements, core pastes, binders, and plastics to take permanent set at normal air temperatures; as opposed to thermal setting.

AQL — Acceptable quality level. A quality level established on a prearranged system of inspection using samples selected at random.

"As Cast" Condition—Casting without subsequent heat treatment.

Assembly—See Spruing.

Back Draft—See Undercut.

Back-Up Coat—The ceramic slurry of dipcoat which is applied in multiple layers following the first dip coat to provide a ceramic shell of the desired thickness and strength for use as a mold.

Binder—Liquid or solid additive to bond refractory particles.

Blended Master Heat — Previously refined metal of two or more single furnace charges that have been intimately mixed. See Master Heat.

Blister—A shadow blowhole with a thin film of the metal over it appearing on the surface of a casting.

Blowhole—Irregularly shaped cavity with smooth walls produced in a casting when gas, evolved during solidification of the metal, fails to escape and is held in pockets.

Bright Dipping — An acid solution into which castings or other articles are dipped to obtain clean, bright surfaces.

Brinell Hardness — The value of a metal's hardness on an arbitrary scale, determined by measuring the diameter of the impression made by a ball of given diameter applied under a known load. Values are expressed in Brinell hardness numbers (BHN).

Buckle—Bulging of a large, flat face of a casting, caused by the dip coat peeling away from the pattern. Usually appears angular with some finning resulting from dip coat cracking.

Burn-out—Firing a mold at a high temperature to remove pattern material residue.

Can—See Flask.

Caustic Dip—Immersion in a solution of fused sodium hydroxide or equivalent, to clean the surface; or, when working with aluminum alloys, to reveal the macrostructure.

Charge—A given weight of metal introduced into the furnace.

Cire-Perdue Process — See Investment Casting Process.

Cluster—A group of expendable patterns on sprue and runners for casting purposes.

Cold Shut—Lines on casting surface resulting from incomplete fusion of metal streams appearing as flow lines on the casting.

Collapsible Core — A metal insert made in two or more pieces to permit withdrawal from an undercut mold surface.

Cooling Curve—A curve showing the relation between time and temperature during the cooling of a metal sample. Since most phase changes involve evolution or absorption of heat, there may be abrupt changes in the slope of the curve.

Core—1. A metal insert in a die to produce a hole in pattern.
   2. See "Preformed Ceramic Core"

Core Cavity—The cavity produced in a casting by use of a core.

Core Print—Projections attached to a pattern to form recesses in the mold at points where cores are to be supported.

Core Shift—A variation from specified dimensions of a cored section due to a change in position of the core or misalignment of cores in assembling.

Core - Soluble Core—A soluble, usually wax, insert that is used to create the internal geometry to the wax pattern. The pattern wax is injected around the core which is subsequently dissolved away to leave the internal geometry required.

Core Wires or Rods—Reinforcing wires or rods for fragile cores, often preformed into special shapes.

Coupon— An extra piece of metal, either cast separately or attached to a casting, used to determine the mechanical or physical properties of the alloy.

Crush—Indentation, usually on the concave face of a casting, resulting from expansion of dip coat into the shell cavity after burnout.

Cut Off — Removing casting from sprue by abrasive wheel or band saw.

Decarburization — Loss of carbon from the surface of a ferrous casting as a result of solidification or heating in a medium, usually oxygen, that reacts with the carbon. Lost carbon can be restored by heat treatment in an appropriate atmosphere.

Degasification—Usually a chemical reaction resulting from a compound added to molten metal to remove gases from the metal. Often inert gases are used in this operation.

Degasifier—A material employed for removing gases from molten metals and alloys.

Dendrite - A crystal that has a treelike branching pattern, being most evident in cast metals slowly cooled through the solidification range.

Deoxidation—Removal of excess oxygen from the molten metal, usually accomplished by adding materials with a high affinity for oxygen.

Deoxidizer—A material used to remove oxygen or oxides from molten metals and alloys.

Descaling—A chemical or mechanical process for removing scale or investment material from castings.

Dewaxing—The process of removing the expendable wax pattern from an investment mold or shell mold, usually accomplished by melting out using steam autoclaves or by flash firing ovens.

Die—The metal form in which the cavity for producing heat-disposable patterns is cut or pressed.

Dip Coat—1. In the solid mold technique of investment casting an extremely fine ceramic coating called pre-coat applied as a slurry directly to the surface of the pattern to reproduce maximum surface smoothness. The coated set up is surrounded by coarser, and more permeable investment to form the mold.

2. In the shell mold technique of investment casting an extremely fine ceramic coating called the first coat, applied as a slurry directly to the surface of the pattern to reproduce maximum surface smoothness. The first coat is followed by other dip coats of different viscosity and usually containing different grading of ceramic particles. After each dip, coarser stucco material is applied to the still wet coating. A build up of several coats forms an investment shell mold.

Directional Solidification—(1) The solidification of molten metals in a casting in such a manner that feed metal always is available for that portion that is just solidifying.

(2) Directional solidification of a casting together with close control of solidification rate so that crystalline orientation of each grain in the casting is parallel to its neighbor, or the production of a single crystal casting.

Discontinuity—An opening in a part representing an actual break in continuity, the term being taken to represent flaws rather than intended openings.

Distortion—Deformation (other than contraction) that develops in a casting between solidification and room temperature, also deformation occurring during heat treating and high temperature service.

Dowel—A pin of various types used in the parting surface of patterns or dies to assure registry.

Ductility—The property permitting permanent (plastic) deformation without rupture in a material when it is stressed in tension.

Ejector Pins—Movable pins in pattern dies which help remove patterns from the die.

Elastic Limit—Maximum stress that a material will withstand without permanent deformation. See Yield Strength.

Electrolytic Etch—Etching an investment casting in an electrically conducting bath as a treatment before further processing or before inspection (see Etching).

Elongation—Amount of permanent extension in the vicinity of the fractures in the tensile test, usually expressed as a percentage of original gage length, such as 25 per cent in 1 in.

Embrittlement—Loss of ductility. See also Hydrogen Embrittlement.

Endurance Limit—A limiting stress, below which the metal will withstand, without rupture, an indefinitely large number of cycles of stress.

Endurance Ratio—The ratio of endurance limit to ultimate strength. For most ferrous metals, this value lies between 0.4 to 0.6. Endurance ratio equals endurance limit divided by ultimate strength.

Etching—In metallography, the process of revealing structural details by preferential attack of reagents on a metal surface. Employed in production of some investment castings to reveal surface grain condition.

Expendable Pattern Material—See Heat Disposable Pattern.

Fatigue—Tendency for a metal to break under conditions of repeated cyclic stressing considerably below the ultimate tensile strength.

Fatigue Crack or Failure—A fracture starting from a nucleus where there is an abnormal concentration of cyclic stress and propagating through the metal. The fracture surface is smooth and frequently shows concentric (sea shell) markings with a nucleus as a center.

Fatigue Strength—Maximum stress that a metal will withstand without failure for a specified large number of cycles of stress. Often confused with endurance limit, to which it is related.

Feeder, Feeder Head—A reservoir of molten metal to compensate for the contraction of metal as it solidifies, thus preventing voids in the casting.

Fillet—Concave corner piece usually used at the intersection of casting sections. Also the radius of metal at such junctions as opposed to an abrupt angular junction.

Fin—A thin projection of metal from the casting, formed either as a result of cracked refractory or from parting line flash which was not removed from the pattern before it was dipped.

Finish Allowance—Amount of stock left on the surface of a casting for machining.

Finish Mark— A symbol (f, fl, f2, etc.) appearing on the line of a drawing that represents the edge of the surface of the casting to be machined or otherwise finished.

Finishing Operation — Machining operations performed on castings after they have been cleaned.

Fired Mold—Shell mold or solid mold which has been heated to a high temperature and is ready for casting.

Flash Dewax—Sudden application of heat to a shell mold for dewaxing; the object of flash dewaxing is to soften and melt the wax or plastic in contact with the shell so fast that heat cannot penetrate deeply enough into the pattern and cause it to expand, with subsequent danger of cracking the shell.

Flask—Tubular or rectangular metal form without top and without fixed bottom used to hold the refractory forming a mold. Also the complete mold with or without pattern material and with or without cast metal.

Flask Liner—Paper or asbestos sheet placed inside a flask before investment is poured in.

Flask Plate—Wood, metal, rubber, or composition base on which flask is set.

Fluorescent Penetrant Inspection—A non-destructive testing method for detecting surface discontinuities with fluorescent material viewed under black (ultra-violet) light.

Foundry Returns—Metal in the form of gates, sprues, runners, risers, and scrapped castings of known composition returned to the furnace for remelting (Remelt).

Fracture Test—Examination of the surface of a deliberately broken test piece or casting to determine structure of the metal or to indicate certain of its properties.

Gaging—Checking dimensional requirements with a gage.

Gas Holes—(Same as Blow Holes) Rounded cavities, either spherical, flattened, or elongated, in a casting, caused by the generation and/or accumulation of gas or entrapped air during solidification of the casting.

Gas Porosity—Dispersion of fine cavities in the metal resulting from the liberation of gas during solidification.

Gate—End of the runner in a mold where molten metal enters the casting or mold cavity. Sometimes applied to entire assembly of connected channels.

Gated Patterns—One or more patterns with gates or channels attached.

Gating System—The cavities in a mold made by a complete assembly of sprues, runners and gates through which molten metal flows to the pattern cavity. Term also applies to the same portions of the pattern set up and the cast cluster.

Grain Refiner—Any material added to a liquid metal for producing a finer grain size in the subsequent casting.

Grain Size—The average size of the crystals or grains in a metal.

Green Permeability—The ability of unfired investment material to permit passage of gases through its mass.

Grinding—Removing gates and other projections from castings by means of an abrasive grinding wheel.

Grit—Crushed ferrous or synthetic abrasive material in various mesh sizes which is used in abrasive blasting equipment for cleaning of castings.

Grog—Burned refractory material.

Gypsum—A common mineral, mostly hydrated calcium sulphate, used in investment molding.

Hardenability—In a ferrous alloy, the property that determines the depth and distribution of hardness induced by quenching.

Hardening—Any process for increasing the hardness of metal by suitable treatment, usually involving heating and cooling.

Heat—A single furnace charge of metal to be used for pouring directly into mold cavities; a heat may be all or part of a master heat.

Heat Disposable Pattern—A pattern formed from wax-base or plastic-base material which is melted from the mold cavity by application of heat.

Heat Treatment—A combination of heating and cooling operations timed and applied to a metal or alloy in the solid state in a manner which will produce desired properties.

HIPping—Hot Isostatic Pressing: A high pressure high temperature method which minimizes internal porosity not connected to the surface of the casting.

Homogenizing—A process of heat treatment at high temperature intended to eliminate or decrease chemical segregation by diffusion..

Hot Tear—Surface discontinuity or fracture caused by either external loads or internal stresses or a combination of both acting on a casting during solidification, and subsequent contraction at temperatures near the solidus.

Hydrogen Embrittlement—A condition of low ductility resulting from the absorption of hydrogen.

Impact Strength—The resistance a material is capable of developing against impact blows; usually expressed as the foot pounds of energy necessary to break a standard specimen.

Inclusions—Particles of slag, refractory materials, sand or deoxidation products trapped in the casting during solidification.

Investing—The process of pouring the investment slurry into a flask surrounding the pattern to form the mold.

Investment—A flowable mixture or slurry of a graded refractory filler, a binder and a liquid vehicle which when poured around the patterns conforms to their shape and subsequently sets hard to form the investment mold.

Investment Casting Process—A pattern casting process in which a wax or thermoplastic pattern is used. The pattern is invested (surrounded) by a refractory slurry. After the mold is dry, the pattern is melted or burned out of the mold cavity, and molten metal poured into the resulting cavity.

Investment Precoat—See Dip Coat.

Investment Shell—Ceramic mold obtained by alternately dipping pattern set up in dip coat slurry and stuccoing with coarse ceramic particles until the shell of desired thickness is obtained.

Knock Out—Removal of castings from investment material.

Lay—In surface finish, lay is defined as direction of the predominant surface pattern. See ASA B46.1 for standard.

Liquidus—A line on a binary phase diagram or a surface on a ternary phase diagram, representing the temperatures at which freezing begins during cooling, or melting ends during heating under equilibrium conditions.

Lost Wax Process—See Investment Casting Process.

Locating Pad— A projection on a casting to assist in maintaining alignment of the casting for machining operations.

Locating Surface—A casting surface to be used as a basis for measurement in making secondary machining operations.

Macrostructure—Structure of metals as revealed by macroscopic examination.

Magnetic Particle Inspection—A nondestructive testing method for inspecting ferrous metals by magnetizing and covering with iron powder to locate discontinuities.

Master Alloy—An alloy of composition high in one or more alloying elements, which is added in making a melt and which permits closer control of composition than is possible with the addition of the pure metals. Also referred to as rich ahoy.

Master Heat—Previously refined metal of a single furnace charge. See Blended Master Heat.

Master Pattern—The object from which a die can be made, generally a metal model of the part to be cast with process shrinkage added.

Mechanical Properties—Those properties of a material that reveal the elastic and inelastic reaction when force is applied, or that involve the relationship between stress and strain; for example, the modulus of elasticity, tensile strength, and fatigue limit. This term should not be used interchangeably with "physical properties".

Metal Lot—A master heat that has been approved for casting and given a sequential number by the foundry.

Microporosity — Extremely fine porosity caused in castings by shrinkage or gas evolution and apparent on radiographic films as mottling. See ASTM Standard E-192 for allowable Microporosity.

Microshrinkage—Very finely divided porosity resulting from interdendritic shrinkage resolved only by use of the microscope; may be visible on radiographic Elms as mottling.

Microstructure — The structure of polished and etched metal and alloy specimens as revealed by the microscope at magnifications over ten diameters.

Mismatch—Offset condition at parting line caused by misalignment of pattern die sections.

Misrun—A casting not fully formed.

Modulus of Rupture—The ultimate strength or the breaking load per unit area of a specimen tested in torsion or in bending (flexure). In tension, it is the tensile strength

Mold Cavity—The impression in a mold produced by removal of the pattern. It is filled with molten metal to form the casting. Gates and risers are not considered part of the mold cavity.

Padding—The process of adding metal to a cross section of a casting wall, usually extending from a riser, to insure adequate feed to a localized area where a shrink would occur if the added metal were not present.

Parting Line—A line on a pattern or casting corresponding to the separation between adjacent sections of a pattern die.

Pattern—A form of wax-base or plastic base material around which refractory material is placed to make a mold for casting metals.

Pattern Assembly—See Cluster.

Pattern Draft—Taper allowed on vertical fame of a pattern to permit easy withdrawal of pattern from the mold or die.

Pattern Injection —The process of filling the pattern die with expendable material, usually in the liquid or plastic state.

Pattern Layout—Full-sized drawing of a pattern showing its arrangement and structural features.

Patternmaker's Shrinkage—Shrinkage allowance made on all patterns to compensate for the change in dimensions as the solidified casting cools in the mold from freezing temperature of the metal to room temperature. Pattern is made larger by the amount of shrinkage characteristic of the particular metal in the casting and the amount of resulting contraction to be encountered. Rules or scales are available for use.

Permeability—The property of a mold material to allow passage of gases.

Photomicrograph—A photograph of the grain structure of a metal as observed when optically magnified more than 10 diameters. The term micrograph may be used.

Physical Properties—Properties of matter such as density, electrical and thermal conductivity, expansion, and specific heat. This term should not be used interchangeably with "mechanical properties."

Pickle—Chemical or electrochemical removal of surface oxides from metal surfaces.

Pilot or Sample Casting—A casting made from a pattern produced in a production die to check accuracy of dimensions and quality of castings which will be made in quantity.

Pinhole Porosity—Very small holes scattered through a casting, possibly caused by microshrinkage or gas evolution during solidification See ASTM Standard E-192 for allowable Pinhole Porosity.

Pit—A sharp depression in the surface of metal. Sometimes called "chrome pitting" in iron-chromium investment casting alloys, which are prevalent to pitting.

Plastic Pattern — Comparable to a wax pattern but formed by injecting molten polystyrene into a die with high pressure.

Porosity—See Gas Porosity, Microporosity, Pinhole Porosity.

Pouring Basin—The enlarged mouth of the sprue into which the molten metal is poured.

Pre-Coating—See Dip Coat.

Pre-Fill—Also called a wax chill. A solid wax insert placed in a pattern die cavity to reduce volumetric shrinkage as molten wax solidifies following injection into the die.

Preformed Ceramic Core — A preformed refractory aggregate inserted in a wax or plastic pattern to shape the interior of that part of a casting which cannot be shaped by the pattern. Sometimes the wax is injected around the pre-formed core.

Radiography—A nondestructive method of internal examination in which metal objects are exposed to a beam of X-ray or gamma radiation. Differences in thickness, density or absorption, caused by internal defects or inclusions, are apparent in the shadow image either on a fluorescent screen or on photographic film placed behind the object.

Rapid Prototyping—Systems using CAD data to produce a prototype pattern to be processed by investment casting. An alternative process is to produce a ceramic mold using CAD data.

Recovery Rate—Ratio of the number of saleable parts to the total number of parts manufactured, expressed as a percentage.

Reduction in Area—The difference between the original cross sectional area of a tensile test piece and that of the smallest area at the point of fracture. Usually stated as a percentage of the original area.

Register —The accuracy of fit or alignment between mating portions of a pattern die.

Reject Rate—Ratio of the number of parts scrapped to the total number of parts manufactured, expressed as a percentage.

Remelt—**See** Foundry Returns.

Ribs—Sections joining parts of a casting to impart greater rigidity.

Riser—A reservoir of molten metal provided to compensate for internal contraction of the casting as it solidifies (also Feeder or Feeder Head).

Rockwell Hardness—The hardness value of a metal determined by measuring the depth of penetration of a 1/16 inch steel ball ("B" Scale) or a diamond point ("C" Scale) using a specified applied load; a type of penetration hardness.

Runner—The portion of the mold cavity between sprue cavity and ingate; applies to same portions of pattern set up and cast cluster.

Scab— Rough-surfaced surplus of metal on a casting where dipcoat has peeled away from outer shells or investment after pattern burnout.

Scleroscope Hardness Test—A hardness test in which the loss in kinetic energy of a falling metal "tup", absorbed by indentation upon impact of the tup on the metal being tested, is indicated by the height of rebound.

Set-Up—See Cluster.

Shell Mold—See Investment Shell.

Shell Slurry—See Dip Coat.

Shrinkage Defect—Jagged hole or spongy area of a casting, lined with dendrites; generally due to insufficient feeding of molten metal during solidification. Not to be confused with Patternmaker's shrinkage.

Shrinkage Allowance — The total corrections for shrinkage, occurring in the making of an investment casting, applied to the master pattern.

Slag—A fused non-metallic material used to protect molten metal from the air and to extract certain impurities.

Slurry— A flowable mixture of refractory particles suspended in a liquid. See Dip Coat.

Solidus—A line on a binary phase diagram representing the temperature at which freezing ends on cooling or melting begins on heating.

Sprue—The portions of the mold cavity between pouring basin and runner cavity; applies to same portions of pattern set up and cast cluster. In top poured casting the

sprue may also act as riser. Sometimes used as generic term to cover all gates, risers, and runners returned to the melting unit for remelting.

Sprue Button—The pattern used to form the pouring basin.

Spruing—Act of attaching patterns to sprue. Also known as wax assembly.

Stress—The intensity (measured per unit area) of the internal distributed forces or components of force which resist deformation of a body. Stress is measured in force per unit area, for example pounds per square inch (p.s.i.).

Stress Raisers—Factors, such as sharp changes in contour or surface defects, which concentrate stresses locally.

Stress Relieving—A process of reducing residual stresses in a casting by heating the casting to a suitable temperature and holding for a sufficient time. This treatment may be applied to relieve stresses induced by casting, machining, welding, or other processing.

Stuccoing—The application of granular refractory to the wet dipcoat on a heat-disposable pattern or pattern cluster. See Investment Shell.

Suction Pouring—Pouring metal into a mold while a vacuum is being drawn through the permeable mold base.

Tensile Strength—The maximum load in tension which a material will withstand prior to fracture. In the case of ductile materials, fracture is preceded by elongation and consequent reduction in cross sectional area. The maximum stress is reached just prior to the necking down of the test piece. Tensile strength is calculated from the maximum load applied during the test divided by the original cross-sectional area.

Test Bar— Standard specimen bar designed to permit determination of mechanical properties of the metal from which it was poured.

Test Coupon—See Coupon.

Test Lug—A lug cast as part of the casting and later removed for testing purposes.

Tie Bar—A bar of metal connecting two comparatively thin sections of a casting separated by a narrow space. Its purpose is to prevent distortion of the sections. Often it is cut off the finished casting

Tolerance—The amount of allowable deviation.

Tumbling — Cleaning castings by rotating them in a container, usually in the presence of cleaning materials.

Undercut—A recess having an opening smaller than the internal configuration, thereby preventing mechanical removal of a one piece core.

Vacuum Casting— Pouring a mold under a vacuum.

Vacuuming—In investment casting the process of removing entrapped air bubbles from the investment or dip coat slurry before, during and/or after investing or coating by subjecting the mixture to a vacuum.

Vacuum Refining — Melting in a vacuum to remove gaseous contaminants from the metal.

Vent—A small opening or passage in a mold or core to facilitate escape of gases when the mold is poured.

Warpage—See Distortion.

Wax Chill—See Pre-Fill.

Wax Injection Dies — A precision metal die that has been cast over a master pattern or into which a cavity of the desired part configuration has been machined. Molten wax or plastic is injected to the cavity under pressure to form a precise wax pattern.

Wax Pattern — A precise duplicate (with shrinkage allowance) of the casting proper and the gates required, which is formed by injecting molten wax with pressure into a die.

X-ray—See Radiography.

Yield Point—The load per unit of original cross section at which marked increase in deformation occurs without increase in load.

Yield Ratio—The ratio of yield strength to ultimate tensile strength.

Yield Strength—The stress at which a material exhibits a specified limit of permanent strain.